AF739935

Progress in Mathematics
Volume 72

Series Editors
J. Oesterlé
A. Weinstein

Izu Vaisman

Symplectic Geometry and Secondary Characteristic Classes

Springer Science+Business Media, LLC 1987

Izu Vaisman
Department of Mathematics
University of Haifa
Mount Carmel, Haifa 31 999
Israel

Library of Congress Cataloging-in-Publication Data
Vaisman, Izu.
Symplectic geometry and secondary characteristic classes / Izu Vaisman.
p. cm. — (Progress in mathematics;v. 72)
Bibliography: p.
Includes index.
ISBN 978-1-4757-1962-8
1. Geometry, Differential. 2. Characteristic classes. 3. Maslov index. I. Title. II. Series: Progress in mathematics (Boston, Mass.);vol. 72.
QA649.V284 1987
516.3'6—dc19 87-19855

CIP-Kurztitelaufnahme der Deutschen Bibliothek
Vaisman, Izu:
Symplectic geometry and secondary characteristic classes / Izu Vaisman.

(Progress in mathematics ; Vol. 72)
ISBN 978-1-4757-1962-8

NE: GT

© Springer Science+Business Media New York 1987
Originally published by Birkhäuser Boston, in 1987
All rights reserved. No part of this publication may be reproduced, stored in a retrieval system, or transmitted, in any form or by any means, electronic, mechanical, photocopying, recording or otherwise, without prior permission of the copyright owner.

ISBN 978-1-4757-1962-8 ISBN 978-1-4757-1960-4 (eBook)
DOI 10.1007/978-1-4757-1960-4

Text prepared by the author in camera-ready form.

9 8 7 6 5 4 3 2 1

FOREWORD

The present work grew out of a study of the Maslov class (e.g. [37]), which is a fundamental invariant in asymptotic analysis of partial differential equations of quantum physics. One of the many interpretations of this class was given by F. Kamber and Ph. Tondeur [43], and it indicates that the Maslov class is a secondary characteristic class of a complex trivial vector bundle endowed with a real reduction of its structure group. (In the basic paper of V.I. Arnold about the Maslov class [2], it is also pointed out without details that the Maslov class is characteristic in the category of vector bundles mentioned previously.) Accordingly, we wanted to study the whole range of secondary characteristic classes involved in this interpretation, and we gave a short description of the results in [83]. It turned out that a complete exposition of this theory was rather lengthy, and, moreover, I felt that many potential readers would have to use a lot of scattered references in order to find the necessary information from either symplectic geometry or the theory of the secondary characteristic classes. On the other hand, both these subjects are of a much larger interest in differential geometry and topology, and in the applications to physical theories. For all these reasons it seemed to me appropriate to give an exposition of symplectic geometry and of the general and specific theory of secondary characteristic classes (including the Maslov class) under the form of a monograph, which we bring now before the readers.

Our approach to the subject is that of differential geometry. While the motivation for the study of the Maslov class comming from analysis and physics is definitely more important, there is also a geometric motivation for this study, and this geometric motivation also asks for the study of appropriate general secondary characteristic classes. Accordingly, we shall not discuss the asymptotic analysis related to the Maslov class (the interested reader may find it, for instance, in [37]), but we motivate our work by the following geometric problem. Let $\pi\colon E \to M$ be a symplectic vector bundle, and let L_0, L_1 be two Lagrangian subbundles of E. We want to study conditions for L_0 and L_1 to be transversal subbundles i.e., $E = L_0 \oplus L_1$, and to measure nontransversality by corresponding obstructions. The theory which we obtain thereby involves a more general case than the one of Kamber-Tondeur [43], which occurs if the bundle E is trivial. Much more important is the fact that the Lagrangian transversality problem is not only of a geometric interest. Indeed, this problem appears in the study of Hamilton-Jacobi equations, and, therefore, in the study of the equations of quantum physics, and this is a basic reason for the study of the transversality of Lagrangian subbundles. Let us also notice that (as we shall point out at the end) the characteristic classes of the triple (E,L_0,L_1) can also play an important role in Lagrangian cobordism theory [4], which is an important part of Lagrangian differential topology.

Symplectic geometry is the framework for the secondary characteristic classes of a triple of vector bundles (E,L_0,L_1) as described previously. Hence, necessarily, we must develop enough symplectic geometry in our work. But, here again, the importance of the subject transcends by far such particular problems as transversality of Lagrangian subbundles. Indeed, symplectic geometry is the framework of all the geometric structures which involve symplectic vector bundles. We discuss general symplectic vector bundles, and natural symplectic vector bundles that appear on symplectic manifolds and their submanifolds, on contact manifolds, etc. In the same setting, we also study various Lagrangian subbundles. A last section of this chapter is devoted to local and tubular equivalence theorems of symplectic geometry: Darboux-Weinstein, Caratheodory, Lie, etc., and a recent theorem of Marle [56] for submanifolds of constant rank of a symplectic manifold. The second half of the book consists of the single Chapter 4 on transversality obstructions of Lagrangian subbundles. Here, we review connections on principal bundles, and then develop a certain version of the theory of the secondary characteristic classes which is simple and adequate to our purpose, namely the Chern-Simons theory [18] and the connection comparison Bott-Lehmann theory [49]. Among others, we give a simple proof of the derivation formulas of Chern-Simons [18] and Heitsch [39]. Then, we insert a review section of various approaches to the standard Maslov class, and an original section on general Maslov secondary characteristic classes (in dimensions 4h-3), which are the obstructions to the transversality of two Lagrangian subbundles of a symplectic vector bundle [83]. (Other known definitions of "higher order Maslov classes" [29], [85], [6], [66] are for more particular situations.) This section includes the general definitions, computation methods, and properties of the new Maslov classes. The last section, which also has an original character, is devoted to the computation of the general Maslov classes for Lagrangian submanifolds of cotangent bundles, endowed with the canonical symplectic structure, and for Legendrian submanifolds of contangent unit spheres bundles of riemannian manifolds. It generalizes a result of Morvan about the standard Maslov class [62], and it establishes a general method of computation showing that the Maslov classes depend on a generalized second fundamental form, and on the curvature of the Lagrangian (Legendrian) submanifold.

I should like to point out that we had no intention whatsoever to write an exhaustive book or textbook on symplectic geometry and/or on secondary characteristic classes. Our book is just a monograph, directed towards the study of the general Maslov classes seen as transversality obstructions, and the material included is determined by this general purpose. It is also determined, by personal preferences in what concerns the more ample development of symplectic geometry than is needed for Maslov classes.

Let us add a few formal explanations. All the objects encountered in this work are in the C^∞ category, and this convention holds without further notice. The differentiable manifolds will be denoted by M,N,..., and if we write M^n, N^h etc. this means that n,h,... are the dimensions of the respective manifolds. $\wedge^p M$, denotes the space of p-forms on M, and $\wedge M$ is the Grassmann algebra of M. TM is the tangent bundle, T*M is the cotangent bundle, etc. Generally, we try to use more or less standard notation. As usual, $\mathbb{R}$ denotes the real field, $\mathbb{C}$ denotes the complex field, Q the quaternion algebra, $\mathbb{Z}$ the integer ring, $\mathbb{R}^n$ etc. the Euclidean n-space over $\mathbb{R}$ etc., S^n the unit sphere in $\mathbb{R}^{n+1}$ and so on. The end of proofs is marked by the abbreviation Q.e.d.

The bibliographical references are far from being exhaustive as well, but we tried to indicate all our sources. We are particularly indepted to such sources as A. Weinstein [87], V. Guillemin and S. Sternberg [37], R. Deheuvels [25], Chern-Simons [18], Bott [15], D. Lehmann [49], J.M. Morvan [62], and to the many other papers quoted in the bibliography at the end of the book.

TABLE OF CONTENTS

CHAPTER 1

INTRODUCTION AND MOTIVATION

The aim of this chapter is to motivate the study of *symplectic geometry* and, particularly, of the *transversality of Lagrangian subbundles* of a symplectic vector bundle by showing that these subjects are imposed necessarily by partial differential equations and mathematical physics. At the same time, we shall introduce the basic notions of symplectic geometry as they grow out of the study of partial differential equations.

1.1. EQUATIONS OF THE HAMILTON-JACOBI TYPE.

Let us consider a system of partial differential equations (p.d.e.) of the following particular type

$$F_\sigma\left(x^i, \frac{\partial f}{\partial x^j}\right) = 0 , \tag{1.1.1}$$

defined on a domain $D \subseteq \mathbb{R}^n = \{(x^1,\ldots,x^n)/x^i \in \mathbb{R}\}$, with the unknown function f, and where, here and in the sequel, we have $i,j,\ldots = 1,\ldots,n$; $\sigma,\tau,\ldots = 1,\ldots,r$.

Definition 1.1.1. *A p.d.e. system of the form (1.1.1) is called a* ***system of the Hamilton-Jacobi type.***

In solving such a system, it is natural to introduce the new variables $\xi_j = \partial f/\partial x^j$, such that the system becomes $F_\sigma(x^i,\xi_j) = 0$, which can be seen as defining a subset C of the space of the $2n$ variables (x^i,ξ_j). C is called the *characteristic set* of the system (1.1.1). Now, the problem of solving (1.1.1) can be divided into two steps: i) find solutions $\xi_j = \xi_j(x^i)$ in C; ii) find such solutions of i) which are the derivatives of a function f.

Furthermore, let us consider a *change of variables*

$$x^i = x^i(\tilde{x}^j),\ \tilde{x}^j = \tilde{x}^j(x^k) \quad . \tag{1.1.2}$$

Then the chain rule for the derivatives implies

$$\tilde{\xi}_i = \frac{\partial x^h}{\partial \tilde{x}^i}\xi_h \quad , \tag{1.1.3}$$

where the Einstein summation convention is used (and it will always be used in this work). Thereby we recognize that the characteristic set C is in fact a subset of the cotangent bundle T^*D, and we define accordingly

Definition 1.1.2. ***A p.d.e. system of the Hamilton-Jacobi type on a differentiable manifold*** *M is a subset C of its cotangent bundle* T^*M.

For the sake of simplicity, we shall assume that C is a submanifold of T^*M. We set $n = \dim M$, and the variables x^i, ξ_j considered above can be seen now as *local canonical coordinates* on T^*M [90]. Because of physical interpretations M is called a *configuration space* and T^*M is the *phase space* [1].

Now, the previously considered steps for solving the p.d.e. system become: i) look for a cross section $s : M \to T^*M$ situated in C; ii) ask $s = df$ for a function $f : M \to \mathbb{R}$, and it is a fundamental fact that these conditions can be expressed by means of some natural geometric structures that exist on T^*M.

Namely, it follows from (1.1.2) and (1.1.3) that

$$\lambda = \xi_i dx^i \tag{1.1.4}$$

is a well defined global 1-form on T^*M, and it is called the *Liouville form.* Its exterior differential is defined by the 2-form

$$\Omega = -d\lambda = dx^i \wedge d\xi_i \quad , \tag{1.1.5}$$

called the *canonical symplectic structure* of T^*M. Now, the condition of step ii) can be expressed by $s^*\lambda = df$, and for such an s to exist it is necessary that $s^*\Omega = 0$. The form Ω is obviously closed and nondegenerate.

In this manner, the study of p.d.e. leads naturally to the following geometric structures

Definition 1.1.3. *Let V be a 2n-dimensional differentiable manifold. Then a closed nondegenerate 2-form ω on V is called a* ***symplectic differential form,*** *and the pair (V,ω) is a* ***symplectic manifold*** *with the* ***symplectic structure*** *ω. The tangent spaces V_p $(p \in V)$ are endowed with* ***bilinear symplectic*** *(i.e., nondegenerate skew symmetric)* ***forms*** *ω_p, and such a linear space (V_p,ω_p) is called a* ***symplectic vector space.*** *Accordingly, the tangent bundle TV is a* ***symplectic vector bundle,*** *while, generally, a vector bundle $\pi : E \to M$ is* ***symplectic*** *if it is endowed with a differentiable field ω of bilinear symplectic forms of its fibers.*

Furthermore, it is clear that $s : M \to T^*M$ defines an n-dimensional submanifold of T^*M, and Ω induces the zero form on this submanifold, which suggests

Definitions 1.1.4. *A n-dimensional (immersed) submanifold N of a symplectic manifold (V^{2n},ω) is called a* ***Lagrangian submanifold*** *if ω induces the zero form on N. Let $(E \to M,\omega)$ be a symplectic vector bundle of rank 2n. A vector subbundle L of E of rank n is called a* ***Lagrangian subbundle*** *if $\omega|_L = 0$. Particularly, if E is just a symplectic vector space, L is a* ***Lagrangian subspace.***

Examples 1.1.5. *a) If N is a Lagrangian submanifold of the symplectic manifold (V,ω), TN is a Lagrangian subbundle of the symplectic*

vector bundle $TV|_N$. *b) For a cotangent bundle* T^*M*, we may distinguish the* ***vertical subbundle*** $V(T^*M)$ *of the tangent spaces of the fibers of* T^*M*. Since these fibers are given by* x^i = const.*, it follows from (1.1.5) that* V *is a Lagrangian subbundle of* $T(T^*M)$*. Moreover,* V *defines a foliation of* T^*M *whose leaves are Lagrangian submanifolds. Generally, such a foliation of a symplectic manifold is a* ***Lagrangian foliation.***

Now, let us come back to the Hamilton-Jacobi system with the characteristic submanifold $C \subset T^*M$, and to the steps i), ii) of its resolution. Step i) may be decomposed again into two parts: i_1) find Lagrangian submanifolds $N \subset C \subset T^*M$; i_2) decide whether N is defined by a cross section $s: M \to T^*M$.

Remark 1.1.6. *We see from* i_1) *that our problem makes sense only if* $h = \dim C \geq n$.

Proposition 1.1.7. *The Lagrangian submanifold* $N \subset T^*M$ *is defined by a cross section* $s: M \to T^*M$ *iff* $\pi|_N$ *is a diffeomorphism. A necessary condition for* N *to be defined by a cross section* s *is that* N *be transversal to the vertical foliation* $V(T^*M)$. .

Proof. The first assertion is obvious. Then, if $\pi|_N$ is a diffeomorphism with inverse s, we have, for $p \in N$,

$$T_pN = s_*T_{\pi(p)}M = \text{span}\left\{s_* \frac{\partial}{\partial x^i}\right\} .$$

Since π has the local equations $x^i = x^i$, we get

$$s_* \frac{\partial}{\partial x^i} = \frac{\partial}{\partial x^i} + \alpha_{ij} \frac{\partial}{\partial \xi_j} \tag{1.1.6}$$

for some coefficients α_{ij} . On the other hand, we have

$$V = \text{span}\left\{\frac{\partial}{\partial \xi_j}\right\} , \tag{1.1.7}$$

hence N is transversal to V. Q.e.d.

Remark 1.1.8. *Conversely, if* N *is transversal to* V*,* T_pN *has a basis of the form (1.1.6), and this basis is sent by* $\pi|_N$ *to the natural basis of* $T_{\pi(p)}M$. *Hence* $\mathrm{rank}(\pi|_N) = n$, *and* $\pi|_N$ *is a local diffeomorphism.*

The transversality condition of Proposition 1.1.7 has the following interpretation. Let us consider the symplectic vector bundle $T(T^*M)|_N = E$ with the basis N. It has two naturally defined Lagrangian subbundles $L_0 = V(T^*M)|_N$, and $L_1 = TN$. The required transversality means $E = L_0 \oplus L_1$, in which case we say that L_0, L_1 are *transversal Lagrangian subbundles* of E.

Accordingly, the p.d.e. systems of the Hamilton-Jacobi type lead us to the study of the geometric problem of the transversality of Lagrangian subbundles. As we shall see later, the *Maslov classes* will be obstructions to the transversality of two Lagrangian subbundles. Hence their study is clearly motivated above.

Let us emphasize that this motivation is much stronger than it might seem. The reason is that the Hamilton-Jacobi type equations are a basic instrument in the study of the most important equations of mathematical physics.

Indeed, many such equations are of the following general form [58]

$$L(x,\lambda^{-1}D_x)(f(x)) \overset{\mathrm{def}}{=} \sum_{|\alpha|=0}^{m} a_\alpha(x)(\lambda^{-1}D_x)^\alpha(f(x)) = 0 \quad , \tag{1.1.8}$$

where $x \in \mathbb{R}^n$, f is the unknown function, $\lambda > 0$ is a "large parameter", and the following notation is used:

$$\alpha = (\alpha_1,\ldots,\alpha_n) \quad \text{for integers} \quad \alpha_i \geq 0 \quad ,$$

$$|\alpha| = \alpha_1 + \ldots + \alpha_n \quad ,$$

(1.1.9) $$D_x = \left(\frac{1}{\sqrt{-1}}\frac{\partial}{\partial x^1},\ldots,\frac{1}{\sqrt{-1}}\frac{\partial}{\partial x^n}\right) \quad ,$$

$$D_x^\alpha = \left(\frac{1}{\sqrt{-1}}\frac{\partial}{\partial x^1}\right)^{\alpha_1} \cdots \left(\frac{1}{\sqrt{-1}}\frac{\partial}{\partial x^n}\right)^{\alpha_n} \quad .$$

For such equations, it is important to find, at least, *asymptotic solutions* of various orders k as $\lambda \to +\infty$, i.e., functions $f(x,\lambda)$ such that $Lf = 0$ up to terms in $\lambda^{-(k+1)}$. It is established that one should look for such solutions under the form

(1.1.10) $$f(x,\lambda) = \{\exp[\sqrt{-1}\lambda S(x)]\} \sum_{i=0}^{\infty} (\sqrt{-1}\lambda)^{-i} \varphi_i(x) \quad ,$$

where S, φ_i are unknown functions. Now, the first condition is that, after the insertion of (1.1.10) in (1.1.8), the term containing λ^0 vanishes. But a straightforward computation shows that this term is

$$\{\exp[\sqrt{-1}\lambda S(x)]\} L\left(x, \frac{\partial S}{\partial x}\right) \varphi_0 \quad ,$$

and it vanishes iff S satisfies the Hamilton-Jacobi equation

(1.1.11) $$L\left(x, \frac{\partial S}{\partial x}\right) = 0 \quad .$$

(Of course, here $\partial S/\partial x$ is seen as the vector $(\partial S/\partial x^1,\ldots,\partial S/\partial x^n)$, and it replaces $\lambda^{-1}D_x$ in (1.1.8).)

Remark 1.1.9. *In the process of looking for global asymptotic solutions of higher order a fundamental obstruction was discovered by J.B. Keller* [44] *and V.P. Maslov* [57], *and it is known as the* ***Maslov Index*** *or the*

(first) ***Maslov class.*** *(See [57], [58], [61], [37], [51], [41]). This class is also, essentially, the first obstruction to the transversality of relevant Lagrangian subbundles, which explains our general use of the term of* ***Maslov classes*** *in this work. Thereby, the role of the first Maslov class transcends in the most significant manner the above mentioned transversality problem, but this theory of the Maslov class is beyond the limits of the present work.*

1.2. **SOME MORE SYMPLECTIC GEOMETRY.** Let us come back again to the Hamilton-Jacobi system of characteristic manifold $C \subset T^*M$. Our notation is the same as in Section 1.1. We should like to discuss a few more aspects concerning the various steps in the resolution of this system.

Let us start with step ii) (Section 1.1), where we have to ensure the relation $s = df$, where $s : M \to T^*M$ sends M onto a Lagrangian submanifold $N \subset T^*M$ such that $\pi|_N$ is a diffeomorphism. This latter condition ensures that N has local parametric equations of the form $\xi_i = \xi_i(x^j)$, and using (1.1.4) we see that $s = df$ means

$$\iota^*\lambda = d(f \circ s^{-1}) \,, \tag{1.2.1}$$

where λ is the Liouville form of T^*M, and $\iota : N \subseteq T^*M$. Since N is Lagrangian, $d\iota^*\lambda = 0$, and f of (1.2.1) always exists locally. It also exists globally iff the cohomology class β defined by $\iota^*\lambda$ in $H^1(N, \mathbb{R})$ is zero.

The still remaining step i_1) (Section 1.1) is more complicated, and it leads to some further developments of symplectic geometry.

Remember that the problem is to produce Lagrangian submanifolds N^n situated in a given submanifold C^h $(n \leq h)$ of a symplectic manifold (V^{2n}, ω). (For the general discussion we do not need to particularize to $V = T^*M$.) The general method to do this is the so-called *method of characteristics*, and we explain it following [37].

Since the submanifold N is defined in V by local implicit equations $f_i = 0$, it is natural to study functions f on open subsets of V that vanish on N and, therefore, $\iota^* df = 0$, where $\iota : N \subseteq V$ is the immersion of N in V.

The symplectic form ω allows us to transpose this condition into a condition expressed by a vector field.

Definition 1.2.1. *The* ***symplectic gradient*** *of a function* f *is the vector field* X_f *defined by*

$$i(X_f)\omega = df \ . \tag{1.2.2}$$

A vector field which is a symplectic gradient is called a ***Hamiltonian vector field.***

Notice that (1.2.2) has a unique solution X_f since ω is nondegenerate. (In (1.2.2) i denotes the *interior product* [59].)

Now it follows that $\iota^* df = 0$ is equivalent to

$$\omega(X_f, Y) = 0 \tag{1.2.3}$$

at every point of N, and for every Y tangent to N. Since $\operatorname{rank} \omega = 2n$, it is known that we cannot have $n+1$ independent vectors $A_1, \ldots, A_{n+1}$ such that $\omega(A_\alpha, A_\beta) = 0$ for all $\alpha, \beta = 1, \ldots, n+1$ (see details in Chapter 2). Hence, if we let Y of (1.2.3) run through a basis of TN, we see that (1.2.3) is equivalent to the fact that X_f is tangent to N i.e., $\iota^* df = 0$ iff $X_f \in TN$.

Furthermore, assume that $p \in C$, and that there is a Lagrangian submanifold N such that $p \in N \subset C$. Let f be a function defined in a neighbourhood of p in V such that $f|_C = 0$. Then $f|_N = 0$, i.e.,

X_f is tangent to N and, therefore, X_f is also tangent to C. If g is one more such function, X_g will also be tangent to N and to C, and so will be $[X_f, X_g]$. But, let us notice that [59]

$$(1.2.4) \qquad i([X_f, X_g])\omega = L_{X_f} i(X_g)\omega - i(X_g) L_{X_f} \omega =$$

$$= di(X_f) i(X_g)\omega + i(X_f) di(X_g)\omega -$$

$$- i(X_g) di(X_f)\omega - i(X_g) i(X_f) d\omega =$$

$$= d(\omega(X_g, X_f)) .$$

where d means exterior derivative, and L means Lie derivative. This suggests

Definition 1.2.2. *Let* $f,g \in C^{\infty}(V)$ = *the ring of differentiable real functions of the symplectic manifold* V. *Then*

$$(1.2.5) \qquad \{f,g\} = \omega(X_g, X_f) = X_f g = -X_g f$$

*is called the **Poisson bracket** of* f,g.

Proposition 1.2.3. *The Poisson bracket satisfies the following properties:*

$$(1.2.6) \qquad X_{\{f,g\}} = [X_f, X_g] \quad ,$$

$$(1.2.7) \qquad \{f,g\} = - \{g,f\} \quad ,$$

$$(1.2.8) \qquad \{f,\{g,h\}\} + \{g,\{h,f\}\} + \{h,\{f,g\}\} = 0 \; .$$

Proof. (1.2.6) is a consequence of (1.2.4). (1.2.7) follows straightforwardly from the definition of the bracket. Finally, (1.2.8) follows by explicitating $d\omega(X_f, X_g, X_h) = 0$ while using (1.2.5) and (1.2.6). Q.e.d.

Since (1.2.8) is the *Jacobi identity*, we have proven the fundamental fact that $(C^{\infty}(V), \{,\})$ is a Lie algebra, called the *Poisson-Lie algebra* of the symplectic manifold V. It plays a basic role in applications to quantum theory.

Now, the considerations which preceed (1.2.4) imply that, for f,g considered there, one has $\{f,g\}$ constant on N. More precisely, from $f|_C = g|_C = 0$, and from (1.2.5) we get $\{f,g\}|_N = 0$, and particularly $\{f,g\}(p) = 0$.

Definition 1.2.4. *The submanifold* $C \subset V$ *is* ***integrable at*** $p \in C$ *if, for every two functions defined in neighbourhoods of* p *in* V *such that* $f|_C = g|_C = 0$, *one has* $\{f,g\}(p) = 0$. C *is* ***integrable*** *if it is* ***integrable*** *at all* $p \in C$.

Accordingly, what we have proven is

Proposition 1.2.5. *Let* C *be a submanifold of a symplectic manifold* (V,ω). *If for every point* $p \in C$ *there exist Lagrangian submanifolds* N *such that* $p \in N \subset C$, *then* C *is an integrable submanifold of* V.

Therefore, the Poisson bracket allows us to formulate a basic necessary condition for the existence of the solutions of a p.d.e. system of the Hamilton-Jacobi type.

If $h = \dim C = n$, then C is integrable iff C is Lagrangian since, in view of (1.2.5) the definitions of these two properties coincide.

Now, let us consider $h > n$, and let $k = 2n-h$ be the codimension of C. Take $p \in C$, and a neighbourhood $U_p \subset V$ such that $U_p \cap C$ is given by the local implicit independent equations

(1.2.9) $$f_1 = 0, \ldots, f_k = 0 .$$

(Assume, for simplicity, that C is an embedded submanifold.) Then, as we know already, $X_{f_1},\dots,X_{f_k}$ are tangent to C, and, since, by the integrability of C, $\{f_a,f_b\}$ are functionally dependent on f_a we must have

$$(1.2.10) \qquad [X_{f_a},X_{f_b}] = \lambda^c_{ab}X_{f_c} \quad ,$$

where $a,b,c = 1,\dots,k$, and λ^c_{ab} are some coefficients. Hence span $\{X_{f_a}\}$ is an integrable distribution, and it defines a foliation of $U_p \cap C$ by k-dimensional leaves.

Remark 1.2.6. *As a matter of fact, the local foliations defined above on neighbourhoods* $U_p \cap C$ *glue up into a global foliation* $\mathcal{F}$ *of* C. *Indeed, if we complete* $\{X_{f_a}\}$ *up to a basis of* TC, *and if we express* $\iota^*\omega$ $(\iota\colon C \subseteq V)$ *by means of this basis, we see that* $\operatorname{rank} \iota^*\omega = h-k = 2(h-n)$, *and* span $\{X_{f_a}\}$ *is the annihilator space of* $\iota^*\omega$. *Hence* span $\{X_{f_a}\}$ *has an invariant meaning.*

Notice, here, the appearance of the geometric structure defined on a manifold C^h by a closed 2-form ω of the constant rank $h-k \leq h$. Such a pair (C,ω) is called a *presymplectic manifold.* On the other hand, ω induces the zero form on the leaves of $\mathcal{F}$. A submanifold of (V,ω) on which ω induces the zero form is called an *isotropic submanifold.* It has a dimension $k \leq n$ (details in Chapter 2), and it is Lagrangian iff $k = n$. Notice that one may define similarly *isotropic vector subspaces* and *isotropic vector subbundles.*

Let us come back to the problem of producing Lagrangian submanifolds in C, and to the previous considerations on the neighbourhood $U_p \cap C$. The method of characteristics is able to produce such a Lagrangian submanifold, locally, whenever one has an $(n-k)$-dimensional isotropic submanifold N_0 such that $p \in N_0 \subseteq C$, and $TN_0 \cap T\mathcal{F} = \{0\}$.

Namely, let us consider the Hamiltonian vector fields X_{f_a} (a=1,...,k) associated to (1.2.9), and assume that the flows $\exp(tX_{f_a})$ are all defined for $|t| < \varepsilon$ with a fixed $0 < \varepsilon \in \mathbb{R}$. Then, the requested Lagrangian submanifold can be obtained by translating N_0 along the flows $\exp(tX_{f_a})$.

More exactly, we shall define

(1.2.11) $$\Phi : N_0 \times (-\varepsilon,\varepsilon)^k \to V$$

by

(1.2.12) $$\Phi(y,(t_1,\dots,t_k)) = \exp(t_k X_{f_k}) \circ \dots \circ \exp(t_1 X_{f_1})(y)$$

for $y \in N_0$. Because of the hypothesis $TN_0 \cap \mathfrak{F} = \{0\}$, Φ defines an n-dimensional submanifold $N \supset N_0$ of V. In order to prove that N is Lagrangian let us establish first

Proposition 1.2.7. *Any Hamiltonian vector field* X_f *on the symplectic manifold* (V,ω) *is an infinitesimal automorphism of the symplectic structure, i.e.,* $\exp(tX_f)$ *preserves the form* ω.

Proof. This fact follows from

(1.2.13) $$L_{X_f}\omega = i(X_f)d\omega + di(X_f)\omega = 0.$$

Q.e.d.

Remark 1.2.8. *Conversely, if one has* $L_X\omega = 0$ *for some vector field* X, *it follows*

$$0 = L_X\omega = i(X)d\omega + di(X)\omega = di(X)\omega \quad .$$

Hence, locally, there is a function f *such that* $i(X)\omega = df$, *and* X *is a* ***locally Hamiltonian vector field.*** *But* X *may not be a (globally) Hamiltonian vector field.*

Now, let us refer to the manifold $N \supset N_0$ constructed by (1.2.12). It is clear that the tangent spaces of N along N_0 are Lagrangian subspaces of TV, and this Lagrangian character is preserved by (1.2.12) because of Proposition 1.2.7. Therefore N is a Lagrangian submanifold of V as desired.

The method of characteristics is of practical value since in the important cases N_0 can be obtained straightforwardly. Indeed, let us assume that $V = T^*M$, and C^h is an integrable submanifold transversal to the vertical foliation $\mathcal{V}(T^*M)$. Let us consider $p_0 \in C$, and local canonical coordinates x^i, ξ_i about p_0 (as in Section 1.1). Because of the transversality to $\mathcal{V}$, C can be defined by local equations of the form

$$(1.2.14) \qquad f_a = \xi_a - \varphi_a(x^1, \ldots, x^n, \xi_{k+1}, \ldots, \xi_n) = 0$$

$$(a = 1, \ldots, k = 2n - h) \ .$$

Checking straightforwardly with (1.2.2) and (1.1.5), we get

$$(1.2.15) \qquad X_{f_a} = \frac{\partial}{\partial x^a} - \sum_{\sigma=k+1}^{n} \frac{\partial \varphi_a}{\partial \xi_\sigma} \frac{\partial}{\partial x^\sigma} + \sum_{i=1}^{n} \frac{\partial \varphi_a}{\partial x^i} \frac{\partial}{\partial \xi_i} \ .$$

Now, it follows that the equations

$$x^1 = x^1(p_0), \ldots, x^k = x^k(p_0) \ ,$$

$$(1.2.16) \qquad \xi_{k+1} = \xi_{k+1}(p_0), \ldots, \xi_n = \xi_n(p_0) \ ,$$

$$\xi_a = \varphi_a\left(x^1(p_0), \ldots, x^k(p_0), x^{k+1}, \ldots, x^n, \xi_{k+1}(p_0), \ldots, \xi_n(p_0)\right),$$

define (locally) the isotropic submanifold N_0 which is needed in order to apply the method of characteristics.

1.3. SOME MORE MATHEMATICAL PHYSICS. In order to bring further motivation for our subject, let us elaborate more about the connection between symplectic geometry and mathematical physics, particularly mechanics.

Let us assume that we have a dynamical system whose *states* are determined by local coordinates q^i $(i = 1,\ldots,n)$, and they describe a differentiable manifold M^n called the *configuration space*. The central problem is to find the trajectories (orbits) of the system by using basic physical principles. We do not insist on these principles here (see, for instance, [3], but recall that one of them is the *principle of minimal action*. It tells us that the system has an associated *Lagrangian function* $L = L(q^i, \dot{q}^i, t)$ (where t denotes time, and $\dot{q}^i = dq^i/dt$), which, physically, is the difference between kinetic and potential energy, and that the trajectories of the system are those who minimize the *action integral* $\int_{t_0}^{t_1} L\, dt$. Accordingly, the trajectories are defined by the *Euler-Lagrange equations* [3]

$$\frac{d}{dt}\left(\frac{\partial L}{\partial \dot{q}^i}\right) - \frac{\partial L}{\partial q^i} = 0 . \tag{1.3.1}$$

Since $(\dot{q}^i)$ can be seen as the coordinates of a tangent vector of M, we may look at (1.3.1) as "living" in the tangent bundle TM, called the *velocity space*, and the trajectories of the system are the projections of the solutions of (1.3.1) in TM onto M.

This viewpoint suggests a reformulation of the equations of motion (1.3.1) which, for the sake of simplicity, we shall describe only locally, and for the case where L does not contain t.

Namely, let p_i denote covector coordinates on M, and let us go over from TM to the cotangent bundle T^*M by means of the so-called *Legendre transformation*

(1.3.2) $$p_i = \frac{\partial L}{\partial \dot{q}^i} \quad , \quad q^i = q^i$$

Now, let us assume that the Lagrangian L is *regular* in the sense that (1.3.2) is a diffeomorphism. Then, we may associate to our dynamical system a new function $H : T^*M \to \mathbb{R}$ called the *Hamiltonian function*, defined by

(1.3.3) $$H = p_i \dot{q}^i - L(q^i, \dot{q}^i) \quad ,$$

where $\dot{q}^i$ are the solutions of (1.3.2).

For H, by using (1.3.2), we get

$$dH = \frac{\partial H}{\partial q^i} dq^i + \frac{\partial H}{\partial p_i} dp_i = \dot{q}^i dp_i + p_i d\dot{q}^i -$$
$$- \frac{\partial L}{\partial q^i} dq^i - \frac{\partial L}{\partial \dot{q}^i} d\dot{q}^i = \dot{q}^i dp_i - \frac{\partial L}{\partial q^i} dq^i \quad ,$$

and it follows from (1.3.1), (1.3.2) that the trajectories of the system are the projections on M of the curves defined in T^*M by *Hamilton's equations*

(1.3.4) $$\frac{dq^i}{dt} = \frac{\partial H}{\partial p_i} \, , \, \frac{dp_i}{dt} = - \frac{\partial H}{\partial q^i} \quad .$$

Hamilton's equations are fundamental in mechanics, p_i are called *momentum variables*, and T^*M is the *phase space* of the system. In the sequel, we shall see that Hamilton's equations have a *symplectic interpretation* in connection with the canonical symplectic structure of T^*M which, in the present notation, is the 2-form

(1.3.5) $$\Omega = dq^i \wedge dp_i \quad .$$

Namely, it follows from (1.2.2), (1.3.5) that the symplectic gradient of the Hamiltonian H is

(1.3.6) $$X_H = \frac{\partial H}{\partial p_i}\frac{\partial}{\partial q^i} - \frac{\partial H}{\partial q^i}\frac{\partial}{\partial p_i}$$

and a comparison with (1.3.4) yields

Theorem 1.3.1. *The solutions of Hamilton's equations are the orbits of the Hamiltonian vector field* X_H *defined by*

(1.3.7) $$i(X_H)\Omega = dH \quad .$$

Therefore, the basic mathematical model of a mechanical system is a symplectic manifold (V,Ω) endowed with a Hamiltonian function H, and (1.3.7) are the global equations of motion. Within this framework it is possible to extend the entire classical theoretical mechanics, and we send to [1] for a full development of this subject. An important role is played by the Poisson bracket. It follows from (1.2.5) and (1.3.5) that

(1.3.8) $$\{f,g\} = \frac{\partial f}{\partial p_i}\frac{\partial g}{\partial q^i} - \frac{\partial f}{\partial q^i}\frac{\partial g}{\partial p_i} \quad ,$$

and Hamilton's equations become

(1.3.9) $$\frac{dq^i}{dt} = \{H,q^i\} \; , \; \frac{\partial p_i}{\partial t} = \{H,p_i\} \quad ,$$

or, globally

(1.3.10) $$X_H f = \{H,f\} \quad ,$$

for any $f : T^*M \to \mathbb{R}$.

The Poisson bracket is also basic for *quantization*. Roughly speaking, the quantization of the mechanical system is a process which associates operators on some Hilbert space with the real functions f on phase space, such as the commutator of two such operators is associated with the Poisson bracket of the functions. (See, for instance, the chapters on quantization in [1] and [37].)

Let us come back for a moment to Hamilton's equations (1.3.4). It is a natural idea to look for solutions of these equations as curves $\gamma(t)$ situated on Lagrangian submanifolds $N \subset T^*M$ defined by cross-sections $dS : M \to T^*M$, where $S : M \to \mathbb{R}$. Locally, such $\gamma(t)$ has the equations

$$q^i = q^i(t) \quad , \quad p_i = \frac{\partial S}{\partial q^i}\left(q^i(t)\right) \quad , \tag{1.3.11}$$

and if we insert them in (1.3.4) we obtain

$$\begin{gathered} \frac{dq^i}{dt} = \frac{\partial H}{\partial p_i}\left(q^i(t) \, , \, \frac{\partial S}{\partial q^i}\left(q^i(t)\right)\right) \quad , \\ \frac{dp_i}{dt} = \frac{\partial^2 S}{\partial q^i \partial q^j}\frac{dq^j}{dt} = \frac{\partial^2 S}{\partial q^i \partial q^j}\frac{\partial H}{\partial p_j} = -\frac{\partial H}{\partial q^i} \quad , \end{gathered} \tag{1.3.12}$$

where, in the second equation, the first two equality signs follow from the second equation (1.3.11), and the first equation (1.3.12).

Now, the second equation (1.3.12) is equivalent to

$$H\left(q^i \, , \, \frac{\partial S}{\partial q^i}\right) = \text{const.}, \tag{1.3.13}$$

which is called the *Hamilton-Jacobi equation of mechanics*. If integrated, it defines S and then $\gamma(t)$ is obtained by the integration of the first equation (1.3.12).

Accordingly, there is a Hamilton-Jacobi equation related to mechanics, and this makes the transversality problem of Lagrangian subbundles important also for theoretical mechanics.

Remark 1.3.2. *For (1.3.13), the method of characteristics leads us back to Hamilton's equations (1.3.4).*

Furthermore, we shall discuss some more problems of mathematical physics, following S. Sternberg [77].

In electrostatics, one has the *electric field strength* E, which, in physics, is seen as a vector field on a domain D of $\mathbb{R}^3$. But, since integration along paths gives the voltage drop, we should rather look at it mathematically as a 1-form E. In this respect, let us remember that the Euclidean metric of $\mathbb{R}^3$ allows for the identification of vectors and covectors.

Remark 1.3.3. *It is a basic idea in* [77] *that the Euclidean character of the metric is imposed by the electric field structure via the dielectric constant.*

It is known that the voltage drop between two points in the domain D does not depend on the path, which means that $\int_\gamma E = 0$ for any closed path γ in D, and hence, E is a closed 1-form, i.e., $dE = 0$ (Stokes' formula). Moreover, if we assume D to be simply connected, we get $E = d\Phi$ for a certain function Φ.

Now, it is possible to study the motion of a particle in D by means of the symplectic-Hamiltonian scheme of mechanics. One has to take the phase space T^*D with coordinates q^i, p_i $(i = 1,2,3)$, and the Hamiltonian function

$$H' = \frac{1}{2m} \sum_{i=1}^{3} p_i^2 + e\Phi(q^i) \quad , \tag{1.3.14}$$

where m is the mass, and e is the electric charge of the particle. This is a modification of the classical Hamiltonian

$$H = \frac{1}{2m} \sum_{i=1}^{3} p_i^2$$

of the motion of a particle in $\mathbb{R}^3$, if no electric field exists. The orbits of the particle, as derived from (1.3.14) and from Hamilton's equations (1.3.4), coincide with those established by experimental physics.

Furthermore, let us assume that we also have on D a time independent magnetic field, defined by its *flux* B which, mathematically, is a 2-form since what it measures are integrals along surfaces in D.

One of the physical laws that govern this structure is that the flux through closed surfaces is zero, i.e., $\int_S B = 0$ for any closed surface $S \subset D$. Equivalently, this means that B is a closed form ($dB = 0$).

Now, if $\pi : T^*D \to D$ is the natural projection, we can learn from experimental physics that the trajectories of the motion of a charged particle of charge e in D are given by Hamilton's equations associated to the Hamiltonian H' of (1.3.14) on the symplectic manifold T^*D endowed with the symplectic form

$$\text{(1.3.15)} \qquad \Omega' = \sum_{i=1}^{3} dq^i \wedge dp_i + e\pi^*B \quad .$$

More generally, if a *spinning particle* is moving in the domain D above, we also need *spinning parameters* to describe it. Accordingly, we shall use the symplectic manifold $V = T^*D \times S^2$, where S^2 is the 2-dimensional unit sphere, and V is endowed with the symplectic form

$$\text{(1.3.16)} \qquad \Omega'' = \Omega' + s\Theta \quad ,$$

where Ω' is defined by (1.3.15), Θ is the area element of S^2, and s is a number called the *spin* of the particle. A modification of the Hamiltonian is also needed in order to refind the experimental results [77].

In the nonstatic case, one has to introduce the general *Maxwell equations* of electromagnetic field on a domain of *space-time* $\mathbb{R}^4$ endowed with the *Minkowski metric*

$$(1.3.17) \qquad ds^2 = c^2dt^2 - \sum_{i=1}^{3} (dq^i)^2 \quad ,$$

where c is the velocity of light.

With the same notation as above, let us introduce the 2-form

$$(1.3.18) \qquad F = B + cE \wedge dt \quad .$$

Then, it can be shown that "half" of the Maxwell equations are equivalent to [77]

$$(1.3.19) \qquad dF = 0 \quad .$$

Furthermore, it follows from experimental physics that the motion of a charged particle, is governed by Hamilton's equation in $T^*\mathbb{R}^4$ with respect to the symplectic form

$$(1.3.20) \qquad \omega = \Omega + e\pi^*F \quad .$$

Moreover, if the particle is spinning, a more complicated symplectic manifold $T^*\mathbb{R}^4 \times \mathcal{O}$, where $\mathcal{O}$ is an orbit of the adjoint representation of the pseudo-orthogonal group, is needed [77].

The interested reader will find in [77] a general construction of a symplectic structure which may be used for the description of the motion of an abstract particle in an abstract *Yang-Mills field* from which the physically interesting cases can be obtained as particular cases. An abstract Yang-Mills field is a principal fibre bundle $P \to M$ of group G, endowed with a connection θ. The abstract particle is represented by means of a sufficiently "nice" symplectic manifold Q on which G acts by *symplectomorphisms* (i.e., diffeomorphisms that preserve the symplectic structure). The symplectic manifold to be used in order to describe the motion of the particle in the field is a *coupled manifold* defined as the total space of an associated bundle $(\pi^{-1}(P) \times Q)/G$, where π is the projection $\pi : T^*M \to M$. The symplectic form contains the curvature of the connection $\pi^{-1}(\theta)$. (See details in [77].)

In closing this chapter, let us point out the existence of important generalizations of the symplectic structure which are still very useful in mathematical physics. One direction is that of infinite dimensional symplectic manifolds with applications in the study of continuous media and field theories [19]. Another direction is the study of manifolds M endowed with a Lie algebra structure on $C^{\infty}(M)$, analogous to the Poisson-Lie algebra, which can be used in mechanics and quantum mechanics [55], [88]. On the other hand, the symplectic structure has important applications in other fields as well, since it gives a framework for general variational calculus [37], But all this is beyond the limits of our work.

CHAPTER 2

SYMPLECTIC VECTOR SPACES

This is a chapter of linear algebra and linear symplectic geometry. We discuss the structure of the symplectic vector spaces, symplectic transformations, and linear subspaces. If the base field is $\mathbb{R}$, we introduce and discuss the compatible complex structures of the symplectic vector space.

2.1. SYMPLECTIC VECTOR SPACES AND THEIR AUTOMORPHISMS. We consider an arbitrary ground field K which is commutative and of characteristic zero, and we formulate again

Definition 2.1.1. *Let* S *be a vector space over* K, *and* $\omega : S \times S \to K$ *be a skewsymmetric bilinear form on* S. *Then* ω *is called a* ***symplectic form*** *or a* ***symplectic scalar product*** *if it is nondegenerate, i.e., the only vector* $v \in S$ *such that* $\forall w \in S$, $\omega(v,w) = 0$ *is* $v = 0$. *A pair* (S,ω) *where* ω *is a symplectic form, is called a* ***symplectic vector space.***

Remark 2.1.2. *Remember that the skew symmetry of* ω *means*

(2.1.1) $$\omega(x,y) = -\omega(y,x) \quad , \quad \omega(x,x) = 0 \quad .$$

Convention 2.1.3. *All the vector spaces considered in this work are finite dimensional.*

Let (S,ω) be a symplectic vector space, and (e_α) be a basis of S. Then we have the coordinate expression

(2.1.2) $$\omega(x,y) = \omega_{\alpha\beta}x^\alpha y^\beta \qquad (x = x^\alpha e_\alpha \ , \ y = y^\beta e_\beta) \quad ,$$

and it is known from elementary linear algebra that ω is nondegenerate iff $\det(\omega_{\alpha\beta}) \neq 0$. Since, by (2.1.1), $(\omega_{\alpha\beta})$ is a skewsymmetric matrix, it follows that the dimension of S must be even. We shall denote $\dim S = 2n$, and agree to the index convention $\alpha, \beta, \ldots = 1, \ldots, 2n$. We shall also always denote

$$\alpha^* = \alpha + n \quad \text{(modulo } 2n). \tag{2.1.3}$$

It is also known from elementary linear algebra that there are bases of S such that (2.1.2) assumes the *canonical form*

$$\omega(x,y) = \sum_{i=1}^{n} (x^i y^{i^*} - x^{i^*} y^i) \quad , \tag{2.1.4}$$

where, following the convention of Chapter 1, $i,j,\ldots = 1,\ldots,n$, and $i^* = i + n$.

If we introduce the *symplectic symbols*

$$s(\alpha) = \begin{cases} -1 & \text{if } \alpha = 1,\ldots,n \\ 1 & \text{if } \alpha = n+1,\ldots,2n \end{cases} \tag{2.1.5}$$

then (2.1.4) is (2.1.2) with the coefficients

$$\omega_{\alpha\beta} = s(\beta)\delta_{\alpha\beta^*} \quad , \tag{2.1.6}$$

δ being the usual Kronecker symbol.

Definition 2.1.4. *A basis* (e_α) *of the symplectic space* (S,ω) *such that (2.1.6) holds good is called a* ***symplectic basis****.*

From the viewpoint of exterior algebra, a symplectic form ω is an exterior 2-form $\omega \in \wedge^2 S^*$, where S^* is the *dual space* of S, and $\wedge$ denotes exterior product. Then, if (e_α) is a symplectic basis and (ε^α) is its dual cobasis, it is easy to check

$$\varepsilon^\alpha = s(\alpha) i(e_{\alpha^*})\omega \tag{2.1.7}$$

(where, as usual, i means interior product), and

(2.1.8) $$\omega = \frac{1}{2} \sum_{\beta=1}^{2n} s(\beta)\varepsilon^{\beta^*} \wedge \varepsilon^{\beta} = \sum_{i=1}^{n} \varepsilon^{i} \wedge \varepsilon^{i^*} \quad .$$

This implies

$$\omega^n = \underbrace{\omega \wedge \ldots \wedge \omega}_{n \text{ times}} = (-1)^{n(n-1)/2} n!\, \varepsilon^1 \wedge \ldots \wedge \varepsilon^{2n} \quad ,$$

and it is usual to define the volume of a parallelotope spaned by 2n vectors by the value of the *volume form*

(2.1.9) $$(-1)^{n(n-1)/2}\omega^n/n! \quad .$$

Then, for a symplectic basis, we get volume 1.

The presence of symplectic scalar product also leads to some other notions similar to those of *Euclidean geometry*.

Definition 2.1.5. *Two vectors* x,y *of the symplectic vector space* (S,ω) *are called* ***skeworthogonal*** *if* $\omega(x,y) = 0$.

By (2.1.1) every vector x is skeworthogonal to itself, hence it is an *isotropic* vector.

We also have analogous of the *musical isomorphisms* of Riemannian geometry. Namely, there is a linear mapping $b : S \to S^*$ given by

(2.1.10) $$b(x) = i(x)\omega \quad ,$$

and, since ω is nondegenerate, b is an isomorphism with the inverse $b^{-1} : S^* \to S$.

These isomorphisms extend to the spaces of tensors and exterior forms of S, and allow for a detailed study of their structure, which is beyond our scope [52].

Now, we shall consider the basic examples of symplectic vector spaces:

a) The space K^{2n} endowed with the symplectic form (2.1.4) is the *standard symplectic space*. It is clear that the choice of a symplectic basis in the symplectic vector space (S,ω) gives an isomorphic identification of the latter with the standard symplectic space.

b) Let T be an arbitrary n-dimensional vector space over K, and T^* be its dual space. Consider $S = T^* \oplus T$ and define $\omega : S \times S \to K$ by

(2.1.11) $$\omega(\tau_1 + t_1 , \tau_2 + t_2) = \tau_1(t_2) - \tau_2(t_1) \quad ,$$

$t_1, t_2 \in T$, $\tau_1, \tau_2 \in T^*$. Then (S,ω) is a symplectic vector space.

It is to be remarked that an arbitrary symplectic space (S,ω) has many decompositions $S = T^* \oplus T$. Indeed, let (e_α) be a symplectic basis of S, and put $T = \text{span}\,\{e_{i*}\}$. Then T^* is isomorphic by b of (2.1.10) with span $\{e_i\}$, and it is easy to check that (S,ω) is isomorphic to $T^* \oplus T$ with the form 2.1.11.

c) Let V be a vector space of dimension $2n+k$ $(k \geq 1)$ over K, and assume that V is endowed with a skewsymmetric bilinear form $\hat{\omega} : V \times V \to K$ which is of *rank 2n*, i.e., if we consider the subspace

(2.1.12) $$F = \{v \in V / \forall w \in V,\ \hat{\omega}(v,w) = 0\} \quad ,$$

called the *annihilator* of $\hat{\omega}$, then $\dim F = k$. Then $\hat{\omega}$ obviously induces a symplectic form ω on $S = V/F$, and (S,ω) is a symplectic vector space. (S,ω) is said to be obtained from $(V,\hat{\omega})$ by *reduction*.

In the remaining part of this Section we shall discuss about the so-called *symplectic group*.

Definition 2.1.6. *Let* $\varphi : (S_1, \omega_1) \to (S_2, \omega_2)$ *be a linear map between two symplectic vector spaces. Then* φ *is said to be a* ***symplectic mapping (transformation)*** *if*

$$(2.1.13) \qquad \omega_2\left(\varphi(x),\varphi(y)\right) = \omega_1(x,y) \qquad (x,y \in S_1) \quad .$$

Since a symplectic form is nondegenerate, a symplectic mapping is necessarily injective, and it is an isomorphism if $\dim S_1 = \dim S_2$. A symplectic isomorphism is also called a *symplectomorphism*. If $(S_1,\omega_1) = (S_2,\omega_2) = (S,\omega)$, φ is an *automorphism* of (S,ω), and it is clear that the set $Sp(S)$ of all these automorphisms, with the usual composition of maps, is a group called the *symplectic group* of (S,ω). Particularly, $Sp(K^{2n}) \overset{\text{def}}{=} Sp(n,K)$ is called the *symplectic linear group in dimension* $2n$ *over* K. Any choice of a symplectic basis in (S,ω) provides a natural identification of $Sp(S)$ with $Sp(n,K)$.

Let us give some more details concerning $Sp(n,K)$. Of course, the elements of $Sp(n,K)$ are identifiable with $(2n,2n)$-nondegenerate matrices (i.e., $Sp(n,K)$ is a subgroup of the general linear group $G\ell(2n,K)$) as follows. If $\varphi \in Sp(n,K)$, it has the coordinate expression

$$(2.1.14) \qquad \tilde{x}^\alpha = \varphi^\alpha_\beta x^\beta \quad ,$$

and we identify $\varphi = (\varphi^\alpha_\beta)$, where the upper index is the line index, and the lower index is the column index. A composition $\psi \circ \varphi$ corresponds to the product matrix $\psi\varphi$. Such matrices φ are called *symplectic matrices*.

It follows from (2.1.6) and (2.1.13) that φ is a symplectic matrix iff

$$(2.1.15) \qquad \sum_{\beta=1}^{2n} s(\beta)\varphi^{\beta^*}_\lambda \varphi^\beta_\mu = s(\mu)\delta_{\lambda\mu^*} \quad .$$

Particularly, we see that the matrix

$$\varphi_0 = \begin{pmatrix} 0 & I_n \\ -I_n & 0 \end{pmatrix} \quad , \tag{2.1.16}$$

where I_n is the (n,n)-unit matrix, is symplectic. This matrix allows us to write the canonical symplectic form (2.1.4) as

$$\omega(x,y) = {}^t x \varphi_0 y \quad , \tag{2.1.17}$$

where x,y are column vectors, and t denotes transposition. It also allows us to write (2.1.15) under the matrix form

$$^t\varphi\varphi_0\varphi = \varphi_0 \quad . \tag{2.1.18}$$

A more detailed expression of the condition for a matrix to be symplectic is obtained if we write

$$\varphi = \begin{pmatrix} A & B \\ C & D \end{pmatrix} \quad , \tag{2.1.19}$$

where A,B,C,D are (n,n)-blocks. Then, the explicitation of (2.1.18) yields

Proposition 2.1.7. *The matrix* φ *given by (2.1.19) is a symplectic matrix iff*

$$^tAC = {}^tCA,\ {}^tBD = {}^tDB,\ {}^tAD - {}^tCB = I_n \quad . \tag{2.1.20}$$

Other interesting properties of the symplectic matrices are given by

Proposition 2.1.8. *If* φ *is a symplectic matrix then: a)* $\det \varphi = 1$*; b)* $\lambda^{2n}P(1/\lambda) = P(\lambda)$*, where* $P(\lambda) = \det(\varphi - \lambda I_{2n})$ *is the characteristic polynomial of* φ*; c) if* λ_0 *is an eigenvalue with multiplicity* k *of* φ*, the same is true for* $1/\lambda_0$.

Proof. a) Since φ is symplectic, it preserves the form ω^n of (2.1.9), i.e.,

$$\varepsilon^1\wedge\ldots\wedge\varepsilon^{2n}(\varphi e_1,\ldots,\varphi e_{2n}) = \det\varphi.\varepsilon^1\wedge\ldots\wedge\varepsilon^{2n}(e_1,\ldots,e_{2n}) =$$

$$= \varepsilon^1\wedge\ldots\wedge\varepsilon^{2n}(e_1,\ldots,e_{2n}) ,$$

whence $\det\varphi = 1$.

b) By using (2.1.18), and since $\det\varphi = 1$, we get

$$\det(\varphi-\lambda I_{2n}) = \det({}^t\varphi-\lambda I_{2n}) = \det[\varphi_0^{-1}({}^t\varphi-\lambda I_{2n})\varphi_0] =$$

$$= \det(\varphi^{-1}-\lambda I_{2n}) = \det\varphi^{-1}\det(I_{2n}-\lambda\varphi) =$$

$$= \lambda^{2n}\det\left(\varphi - \frac{1}{\lambda} I_{2n}\right) .$$

c) This is a straightforward consequence of b). Notice that $\lambda_0 \neq 0$ because of part a) above. Q.e.d.

Remark 2.1.9. *If* $K = \mathbb{R}$, *and* φ *has the eigenvalue* λ_0 *it also has the eigenvalues* $\bar{\lambda}_0$, $1/\lambda_0$, $1/\bar{\lambda}_0$. *(The bar denotes complex conjugation.)*

Finally, let us indicate that our main interest will be for the case of the real ground field $K = \mathbb{R}$. Then, it is clear that the *real symplectic group* $Sp(n,\mathbb{R})$ is a Lie group, and, by standard methods [59], we deduce that its Lie algebra, denoted $sp(n,\mathbb{R})$ is the algebra of $(2n,2n)$-matrices h such that

$$\omega(hx,y) + \omega(x,hy) = 0 \quad , \tag{2.1.21}$$

where ω is the canonical symplectic form on $\mathbb{R}^{2n}$. In matrix form, this condition is equivalent to

$${}^t h\varphi_0 + \varphi_0 h = 0 \quad . \tag{2.1.22}$$

Remark 2.1.10. *(2.1.22) yields* $^t h = -\varphi_0 h \varphi_0^{-1} = \varphi_0 h \varphi_0$ *, whence*

$$\det(h-\lambda I_{2n}) = \det(\varphi_0 h \varphi_0^{-1} - \lambda I_{2n}) = \det(-{}^t h - \lambda I_{2n}) =$$

$$= \det(h+\lambda I_{2n}) \quad .$$

Accordingly, h *has the eigenvalues* λ *and* $-\lambda$ *simultaneously and with the same multiplicities.*

The structure of the symplectic automorphisms can be further analyzed by means of a very interesting decomposition as a product of so-called *transvections*. The latter play in symplectic geometry the role which reflections play in Euclidean geometry. We refer the interested reader, for instance, to [25].

2.2. SUBSPACES OF SYMPLECTIC VECTOR SPACES. Let A be a k-dimensional linear subspace of the symplectic vector space (S^{2n},ω) over the ground field K. Then k is invariant by symplectic automorphisms of S. Another natural invariant is rank ω_A, where $\omega_A \overset{\text{def}}{=} \omega/_A$, which is an even number $2\ell \leq k$. We shall prove that k and 2ℓ classify the subspaces of S^{2n} since they are the only independent symplectic invariants of subspaces.

Definition 2.2.1. *Given the subspace* $A \subseteq S$, *the subspace*

$$(2.2.1) \qquad {}^s A = \{x \in S/\forall y \in A, \omega(x,y) = 0\}$$

is called the ***skeworthogonal subspace*** *of* A.

Clearly, dim ${}^s A$ = 2n-k, but, unlike the Euclidean case we may have

$$(2.2.2) \qquad {}^i A \overset{\text{def}}{=} A \cap {}^s A = \{x \in A/i(x)\omega_A = 0\} \neq 0 \quad .$$

For instance, if k = 1, $\omega(x,x) = 0$ implies ${}^i A = A \subset {}^s A$. Generally definition (2.2.2) yields

$$\dim {}^{i}A = k - 2\ell \quad . \tag{2.2.3}$$

We call ${}^{i}A$ the *radical* of the subspace A.

A few other properties of (2.2.1) are contained in

Proposition 2.2.2. *a)* $A \subset B$ *implies* ${}^{s}B \subset {}^{s}A$; *b)* ${}^{ss}A = A$; *c)* ${}^{s}(A+B) = {}^{s}A \cap {}^{s}B$; *d)* ${}^{s}(A \cap B) = {}^{s}A + {}^{s}B$.

The proof is straightforward (+ denotes union of subspaces).

Another important subspace associated to A is

$${}^{u}A = A + {}^{s}A. \tag{2.2.4}$$

We call it the *union subspace* of A, and, by a well known formula for dimensions, we have $\dim {}^{u}A = 2n - (k-2\ell)$.

Finally, by example c) of Section 2.1, and since ${}^{i}A$ is the annihilator of ω_A, we have an associated *reduced* symplectic space

$${}^{r}A = A/{}^{i}A \qquad (\dim {}^{r}A = 2\ell). \tag{2.2.5}$$

In the sequel, we make a choice of a supplementary subspace B such that $A = {}^{i}A \oplus B$, and it is clear that (B,ω_B) is a symplectic space isomorphic to ${}^{r}A$.

Similarly, if we choose C such that ${}^{s}A = {}^{i}A \oplus C$, then (C,ω_C) is a symplectic space isomorphic to the reduced space ${}^{rs}A = {}^{s}A/{}^{i}A$, and $\dim C = 2(n-k+\ell)$.

Of course, the choice of B and C is not unique.

Before proceeding, let us formulate

Definitions 2.2.3. *A subspace* Q *of* rank 0 *is called an* ***isotropic subspace*** *of* (S,ω). *(Equivalently* $\omega/_Q \equiv 0$.*) A subspace* $P \subset S$ *such that* $\omega/_P$ *is nondegenerate is called a* ***symplectic subspace*** *of* S.

Now, we can prove the following basic result

Theorem 2.2.4. *Given the subspace* $A \subset S$, *there is a (non-unique) isotropic subspace* $Q \subset S$, *supplementary to* ${}^uA$, *such that* ${}^\sigma A \overset{\text{def}}{=} {}^iA \oplus Q$ *is a symplectic subspace of* S, *and*

$$(2.2.6) \qquad S = {}^\sigma A \oplus_s B \oplus_s C \quad ,$$

where $\oplus_s$ *denotes a direct sum of skeworthogonal terms.*

Proof. Clearly, ${}^s(B \oplus C) \supset {}^iA$, and we may choose a subspace Q' such that ${}^iA \oplus Q' = {}^s(B \oplus C)$. Necessarily, Q' is supplementary to ${}^uA$, and $\dim Q' = \dim {}^iA$.

Furthermore, it is easy to check that the isomorphism b defined by (2.1.10) induces an isomorphism

$$(2.2.7) \qquad {}^iA \underset{(b)}{\approx} (S/{}^uA)^* \approx Q'^* \quad .$$

Hence, if (a_t) $(t = 1,\ldots,k-2\ell)$ is a basis of ${}^iA$, $b(a_t)$ defines a basis of Q'^*, and, thereby, a dual basis q'_t in Q' such that $\omega(a_s,q'_t) = \delta_{st}$ $(s,t = 1,\ldots,k-2\ell)$.

Now, let us consider the transformation

$$(2.2.8) \qquad q_t = q'_t + \lambda_t^s a_s \quad ,$$

and fix the coefficients λ_t^s such that $\omega(q_s,q_t) = 0$, which is clearly possible. Then, it is easy to check that the subspace $Q = \text{span}\,\{q_t\}$ is isotropic, and it satisfies all the requirements of Theorem 2.2.4. It also satisfies the duality relation (2.2.7), i.e., $A \approx Q^*$. Q.e.d.

Remark 2.2.5. *The coefficients* λ_t^s *of (2.2.8) may be determined by the conditions* $\lambda_t^s = -\lambda_s^t$ *, and* $\lambda_t^s = (1/2)\omega(q'_s, q'_t)$. *Then, it follows by a straightforward computation that if the basis* a_s *is replaced by another basis* $\tilde{a}_t = \alpha_t^s a_s$, *the space* Q *remains unchanged.*

Definition 2.2.6. *A symplectic basis* (e_α) $(\alpha = 1,\ldots,2n)$ *of* S *is said to be* ***adapted to*** A *if: i)* (e_σ, e_{σ^*}) $(\sigma = 1,\ldots,k-2\ell)$ *is a basis of* ${}^\sigma A$ *with* $e_\sigma \in {}^1A$, $e_{\sigma^*} \in Q$; *ii)* (e_τ, e_{τ^*}) $(\tau = k-2\ell+1,\ldots,k-\ell)$ *is a basis of* B; *iii)* (e_ξ, e_{ξ^*}) $(\xi = k-\ell+1,\ldots,n)$ *is a basis of* C.

Of course, the adaptedness of a basis depends on the choice of the decomposition (2.2.6). The existence of adapted bases is obvious.

Corollary 2.2.7. *If* A_1, A_2 *are subspaces of* (S,ω) *of the same dimension* k, *and the same* rank 2ℓ, *there exists a symplectic automorphism* σ *of* S *such that* $\sigma(A_1) = A_2$.

Proof. Take two symplectic bases adapted to A_1, A_2, respectively, and take for σ the automorphism which sends the first basis into the second one. Q.e.d.

This corollary precisely means that dimension and rank are the only independent symplectic invariants of a subspace.

Corollary 2.2.8. *(Witt's theorem). Let* $(S_a, \omega_a)_{(a=1,2)}$ *be two symplectic vector spaces of the same dimension* 2n, *and* $A_a \subset S_a$ *two subspaces of the same dimension* k. *Let* $\varphi : A_1 \to A_2$ *be a linear isomorphism such that*

$$(2.2.9) \qquad (\forall x,y \in A_1) \quad \left[\omega_2(\varphi x, \varphi y) = \omega_1(x,y)\right]$$

Then φ *can be extended to a symplectic isomorphism* $\tilde{\varphi} : S_1 \to S_2$.

Proof. Because of (2.2.9), A_a $(a = 1,2)$ have also the same rank. Then, we may take adapted symplectic bases whose parts in A_1, A_2 correspond to each other by φ, and define $\tilde{\varphi}$ as the symplectic isomorphism which sends the first basis onto the second one. Q.e.d.

Now, let us supplement Definition 2.2.3 by

Definitions 2.2.9. *A subspace* $A \subset (S,\omega)$ *is said to be* ***coisotropic*** *if* ${}^{s}A$ *is isotropic. A subspace* $L \subset (S,\omega)$ *which is both isotropic and coisotropic is called a* ***Lagrangian subspace.***

Definitions 2.2.3 and 2.2.9 introduce the most important types of subspaces, and we shall indicate some more properties of these subspaces. First, it follows from (2.2.2), (2.2.3) that A is isotropic iff $A \subset {}^{s}A$, hence A is coisotropic iff ${}^{s}A \subset A$, and A is Lagrangian iff $A = {}^{s}A$. On the other hand, A is symplectic iff ${}^{i}A = \{0\}$.

Moreover, if A is isotropic, $k \leqslant 2n-k$, i.e., $k \leqslant n$, and iff A is coisotropic $k \geqslant n$. Hence, a Lagrangian subspace is n-dimensional, and, coversely, an n-dimensional isotropic subspace is Lagrangian. If A is isotropic, S has decompositions (2.2.6) where $B = \{0\}$, and the vectors e_i of an adapted symplectic basis span a Lagrangian space $L \supset A$. Now, it follows easily that a Lagrangian subspace can also be defined as a maximal isotropic subspace of S. Finally, if A is isotropic, it is Lagrangian in the symplectic subspace ${}^{\sigma}A$ of (2.2.6), and if A is Lagrangian, we have $S = {}^{\sigma}A \approx A \oplus A^*$.

Remark 2.2.10. *A hyperplane* P *of* S *is necessarily coisotropic. Indeed, since* $k = 2n-1$, $\dim {}^{s}P = 1$, *and* $\operatorname{rank} {}^{s}P = 0$.

The following Proposition gives a remarkable property of Lagrangian and coisotropic subspaces [1], [87]

Proposition 2.2.11. *Let* $L \subset (S,\omega)$ *be a Lagrangian subspace, and* $A \subset (S,\omega)$ *be a coisotropic subspace. Then the image of* $L \cap A$ *by projection into the reduced space* ${}^{r}A$ *is a Lagrangian subspace.*

Proof. Since A is coisotropic we have ${}^{r}A/{}^{s}A$, and for any subspace $T \subset S$ we may identify the image of $T \cap A$ in ${}^{r}A$, denoted by $T \cap A \pmod{{}^{s}A}$, with $T \cap A / T \cap {}^{s}A$. Hence

$$\dim \{T \cap A \pmod{{}^{s}A}\} = \dim (T \cap A) - \dim (T \cap {}^{s}A).$$

We claim that ${}^{s}\{T \cap A \pmod{{}^{s}A}\} = {}^{s}T \cap A \pmod{{}^{s}A}$ in ${}^{r}A$. Indeed, these spaces are skeworthogonal, and we have

$$\begin{aligned} &\dim \{T \cap A \pmod{{}^{s}A}\} + \dim \{{}^{s}T \cap A \pmod{{}^{s}A}\} = \\ &= \dim (T \cap A) - \dim (T \cap {}^{s}A) + \dim ({}^{s}T \cap A) - \dim ({}^{s}T \cap {}^{s}A) = \\ &= \dim T + \dim A - \dim (T + A) - \dim T - \dim {}^{s}A + \\ &+ \dim (T + {}^{s}A) + \dim ({}^{s}T \cap A) - 2n + \dim (T + A) = \\ &= \dim A - \dim {}^{s}A = \dim {}^{r}A. \end{aligned}$$

For $T=L$ of our Proposition, this gives

$${}^{s}\{L \cap A \pmod{{}^{s}A}\} = {}^{s}L \cap A \pmod{{}^{s}A} = L \cap A \pmod{{}^{s}A}. \quad \text{Q.e.d.}$$

Remark 2.2.12. *There are counterexamples showing that the result of Proposition 2.2.11 may not hold if* A *is not coisotropic.*

Now, let us consider the set $\mathcal{L}(S)$ of all the Lagrangian subspaces of (S,ω). It follows from Corollary 2.2.7 that the symplectic group $Sp(S)$ acts transitively on $\mathcal{L}(S)$, and the latter is a homogeneous space [59] $Sp(S)/G_L$, G_L being the isotropy subgroup of symplectic automorphisms that preserve a fixed Lagrangian subspace L.

Let (e_i, e_{i*}) $(i=1,\ldots,n)$ be an adapted symplectic basis of L (i.e., $e_i \in L$). Then S is identified with K^{2n}, $Sp(S)$ with $Sp(n,K)$, and L with the subspace L_0 spanned by the first n vectors of the natural basis of K^{2n}. Accordingly, $G_L \approx G_{L_0}$ which is the group of symplectic matrices of the form

$$\begin{pmatrix} A & B \\ 0 & D \end{pmatrix} . \tag{2.2.10}$$

where, like in (2.1.19] the matrix is composed by (n,n)-blocks (here C of (2.1.19) is the zero matrix), and in view of (2.1.20) we have

(2.2.11) $\quad {}^tBD = {}^tDB, \quad {}^tAD = I_n \qquad (\text{i.e., } D = ({}^tA)^{-1})$.

Hence we have proven

Proposition 2.2.13. $\mathcal{L}(S) \approx Sp(n,K^{2n})/G_{L_0}$.

Furthermore, for the fixed Lagrangian subspace $L \subset S$, we have the set $T(L) = \{L' \in \mathcal{L}(S)/L \oplus L' = S\}$, i.e., of Lagrangian subspaces L' *transversal to* L. From Theorem 2.2.4, and using adapted frames of L, it follows easily that $T(L)$ is nonvoid, and that G_L acts transitively on $T(L)$. If we fix $L' \in T(L)$, and an adapted basis $e_i \in L$, $e_{i*} \in L'$, the isotropy subgroup $G_{L,L'}$ of the symplectomorphisms leaving the pair (L.L') invariant will be identified with the group of the matrices

(2.2.12) $$\begin{pmatrix} A & 0 \\ 0 & {}^tA^{-1} \end{pmatrix}$$

as it follows immediately from (2.2.11). This group is isomorphic with the general linear group $G\ell(n,K)$, hence

Proposition 2.2.14. $T(L) \approx G_{L_0}/G\ell(n,K)$.

As a consequence, we shall establish the more interesting result

Theorem 2.2.15. T(L) *has a natural structure of an* n(n+1)/2-*dimensional affine space over* K.

Proof. Remember that an *affine space* is a triple (A,V,π) where A is a set, V a vector space, and $\pi: A \times A \to V$ is a mapping such that for a fixed *point* $a_0 \in A$, $\pi/\{a_0\}\times A$ is a bijection, and for any triple a,b,c $\{a,b,c\} \subset A$ one has $\pi(a,b) + \pi(b,c) = \pi(a,c)$. The *dimension* of the affine space is dim V, and a_0 is an *origine* [10].

Now, let us fix $L' \in \mathcal{T}(L)$, and an adapted basis such as to have the identification of Proposition 2.2.14. Then, $L'' \in \mathcal{T}(L)$ becomes a class of matrices (2.2.10) modulo (2.2.12), defined by the fact that the respective symplectic transformations leave L invariant, and send L' onto L''. By multiplying (2.2.10) with (2.2.12) where A is replaced by A^{-1}, we get a matrix of the form

$$(2.2.13) \qquad \begin{pmatrix} I_n & M \\ 0 & I_n \end{pmatrix}$$

where $M = {}^tM$ is a symmetric (n,n) matrix and it is easy to see that a class of $G_{L_0}/G\ell(n,K)$ has one and only one representative of the form (2.2.13). This representative is the matrix of the symplectic transformation σ which leaves invariant all the vectors of L and sends L' to L''.

The symplectic transformation σ has an invariant meaning, while the matrix M of (2.2.13) is obtained by expressing σ by means of a choosen adapted basis (e_i, e_{i*}) as above. If this basis is changed by $\tilde{e} = e\,A$, $e_* = \tilde{e}_*\,{}^tA^{-1}$, where $e, \tilde{e}, e_*, \tilde{e}_*$ are $(1,n)$-matrices, and A is a nonsingular (n,n)-matrix (remember also the b-duality relation of L, L'), then it follows easily that M is replaced by the matrix $A^{-1}M\,{}^tA^{-1}$. Hence the matrix M is associated with an invariant quadratic form $Q(L',L'')$ on the linear space L^*. If L^* is identified to L', then $Q(L',L'') = \omega(\sigma(\cdot),\cdot)$ for arguments in L'.

Now, let Ξ_L denote the set of quadratic forms on L^*. It is easy to see that Ξ_L has a natural structure of a $n(n+1)/2$-dimensional vector space over K. The dimension is obtained by using the representative matrices M. The use of the same matrices (i.e., the use of a chosen adapted basis) also allows for a straightforward checking of the relation

$$Q(L',L'') + Q(L'',L''') = Q(L',L''')$$

for $L',L'',L''' \in \mathcal{T}(L)$. Hence the triple $(\mathcal{T}(L), \Xi_L, Q : \mathcal{T}(L) \times \mathcal{T}(L) \to \Xi_L)$ is an affine space. Q.e.d.

Remark 2.2.16. *If* $L',L'' \in \mathcal{T}(L)$ *then* L' *and* L'' *are transversal Lagrangian subspaces iff* $Q(L',L'')$ *is a nondegenerate form. We see this by looking at the bases of* L,L',L'', *and at (2.2.13).*

Proposition 2.2.17. *For any pair* (L,L') *of Lagrangian subspaces of* (S,ω), *there exist common transversal Lagrangian subspaces* L''.

Proof. If L and L' are transversal, we get L'' by means of Remark 2.2.16. If not, we shall remark that

$$^s(L+L') = {^sL} \cap {^sL'} = L \cap L' \subset L + L'$$

i.e., $L + L'$ is a coisotropic subspace, and it has the reduced symplectic space $^r(L+L') = (L+L')/L \cap L'$. By Proposition 2.2.11, L and L' project to the transversal Lagrangian subspaces $\tilde{L},\tilde{L}' \subset {^r(L+L')}$. Therefore, there is a Lagrangian subspace $\tilde{L}'' \subset {^r(L+L')}$ which is transversal to $\tilde{L},\tilde{L}'$. The lift $\bar{L}''$ of $\tilde{L}''$ to $L + L'$ is easily seen to be a Lagrangian subspace of S whose intersection with L, L' is $L \cap L'$, and we may choose a supplementary subspace T such that $(L \cap L') \oplus T = \bar{L}''$. Now, let us apply Theorem 2.2.4 for $A = L+L'$, $^1A = {^sA} = L \cap L'$, and $B \supset T$. Then, we have the corresponding space Q of that Theorem, and it is clear that $L'' = T \oplus Q$ is a Lagrangian subspace of S, transversal to both L and L'. Q.e.d.

Proposition 2.2.18. [21]. *Let* L,L',L'' *be Lagrangian subspaces of* (S,ω) *such that* $L \cap L' = L \cap L''$. *Then there is a (nonunique) symplectic transformation of* S *which leaves every vector of* L *invariant and sends* L' *onto* L''.

Proof. Since the involved subspaces are Lagrangian, the subspace

$$A = {}^{s}(L \cap L') = {}^{s}(L \cap L'') = L + L' = L + L''$$

is coisotropic with the radical $L \cap L'$ of dimension $k \leq n$ and with the reduced symplectic space ${}^{r}A = A/L \cap L'$ of dimension $2(n-k)$.

Now, L, L', L'' project to $\tilde{L}, \tilde{L}', \tilde{L}''$ which are Lagrangian subspaces of ${}^{r}A$, and $\tilde{L}', \tilde{L}''$ are transversal to $\tilde{L}$. Let us denote the elements of ${}^{r}A$ by brackets of vectors of A, and let $[e_{k+1}], \ldots, [e_n]$ be a basis of $\tilde{L}$ with the dual $[e_{(k+1)*}], \ldots, [e_{n*}]$ in $\tilde{L}'$. Then, the symplectomorphism σ of ${}^{r}A$ which fixes every element of $\tilde{L}$, and sends $\tilde{L}'$ to $\tilde{L}''$ is given by

$$[e_\sigma] \mapsto [e_\sigma], \quad [e_{\sigma*}] \mapsto m^{\tau}_{\sigma*}[e_\tau] + [e_{\sigma*}] \quad , \tag{2.2.14}$$

where $\sigma, \tau = k+1, \ldots, n$, $\sigma* = \sigma+n$, and $m^{\tau}_{\sigma*}$ is a symmetric matrix (see the proof of Theorem 2.2.15). Of course, the choice of the representative vectors $e_\sigma, e_{\sigma*}$ is not unique.

Furthermore, let $e_1, \ldots, e_k$ be a basis of $L \cap L'$, and $e_{1*}, \ldots, e_{k*}$ be its b-dual basis in the space Q associated to our subspace A by Theorem 2.2.4. Then, (e_i, e_{i*}) $(i = 1, \ldots, n)$ is a symplectic basis in S, and we may define a symplectic transformation by

$$\begin{aligned} & e_i \mapsto e_i \quad (i = 1, \ldots, n); \quad e_{a*} \mapsto e_{a*} \quad (a = 1, \ldots, k) \;; \\ & e_{\sigma*} \mapsto m^{\tau}_{\sigma*} e_\tau + e_{\sigma*} \qquad\qquad (\sigma = k+1, \ldots, n), \end{aligned} \tag{2.2.15}$$

where m is the matrix of (2.2.14). This is precisely a transformation as required. Q.e.d.

Remark 2.3.19. *The coordinate expression of the transformation (2.2.15) has again a matrix of the form (2.2.13) but now one has*

$$(2.2.16) \qquad M = \begin{pmatrix} 0 & 0 \\ 0 & m \end{pmatrix} \quad .$$

2.3. COMPLEX STRUCTURES IN REAL SYMPLECTIC SPACES. In this Section, the ground field is $K = \mathbb{R}$, without any further notice. The basic supplementary fact, now, is the existence of compatible complex structures. We assume that the basic facts of Hermitian geometry are known [25], [31], [45].

Definition 2.3.1. *Let* (S^{2n},ω) *be a real symplectic vector space, and* $J : S \to S$, $J^2 = -\mathrm{Id}$, *be a complex structure on* S. *Then* J *is said to be* ***compatible*** *with the symplectic structure* ω *if*

$$(2.3.1) \qquad \omega(Jx,Jy) = \omega(x,y) \quad .$$

Let us remember that (S,J) is a complex vector space where $\sqrt{-1}\, v = J\, v$. If $S_c = S \otimes_{\mathbb{R}} \mathbb{C}$ ($\mathbb{C}$ is the complex field) is the ***complexification*** of S, J extends by linearity to S_c, where it has the proper values $\pm\sqrt{-1}$, and two n-dimensional proper subspaces

$$(2.3.2) \qquad S' = \{v - \sqrt{-1}\, Jv / v \in S\}, \quad \overline{S}' = \{v + \sqrt{-1}\, Jv / v \in S\} \quad ,$$

such that $S_c = S' \oplus \overline{S}'$, and the bar denotes complex conjugation. Now, the complex n-dimensional vector space (S,J), can be identified with S', and usual multiplication by $\sqrt{-1}$.

Let J be a compatible complex structure of (S,ω), and, for $v,w \in S$, put

$$(2.3.3) \qquad g(v,w) = \omega(v,Jw) \quad (\text{or} \quad g(Jv,w) = \omega(v,w)) \; .$$

Then it follows easily that

$$(2.3.4) \qquad g(v,w) = g(w,v) \quad , \quad g(Jv,Jw) = g(v,w) \; ,$$

and the symmetric bilinear form g is nondegenerate.

Definition 2.3.2. *A form* g *defined by (2.3.3) is called a* ***ω-compatible pseudohermitian metric.*** *If, moreover,* $g(v,v) \geq 0$ *for all* $v \in S$, g *is called a* ***hermitian metric,*** *and* J *is a* ***positive compatible complex structure.***

Theorem 2.3.3. *Every real symplectic vector space* (S,ω) *admits compatible positive complex structures and hermitian structures. If* J_0, J_1 *are two such structures, they are homotopic, i.e., there is a differentiable family* J_t, $0 \leq t \leq 1$, *of positive compatible complex structures on* S *which starts with* J_0 *and ends with* J_1 .

Proof. A very simple way to get J is to consider a pair L, L' of transversal Lagrangian subspaces, and an adapted symplectic basis $e_i \in L$, $e_{i*} \in L'$, then to put

$$(2.3.4) \qquad Je_i = e_{i*} \quad , \quad Je_{i*} = -e_i \quad .$$

The compatibility and positivity of J follow straightforwardly.

A more invariant formulation of this construction is: take the transversal pair of Lagrangian subspaces L, L', and a Euclidean scalar product on L, with the induced isomorphism $c : L \to L^*$; define $J/_L = b^{-1} \circ c$, where b is ω-duality, and $J/_{L'}$ such that $J^2 = -Id$. Then J is the requested complex structure; it sends L to L', and we shall denote it by $J_{(L,L',c)}$.

However, in order to derive the second assertion of Theorem 2.3.3, and to generalize later to vector bundles, we shall give one more proof for the existence of J (e.g., [1]).

Let γ be a Euclidean scalar product on S, and define $a : S \to S$ by

(2.3.5) $\gamma(av,w) = \omega(v,w) \qquad (v,w \in S)$.

Then, using the antisymmetry of ω, we get

(2.3.6) $\gamma(av,w) = -\gamma(v,aw)$, $\gamma(a^2v,w) = -\gamma(av,aw) = \gamma(v,a^2w)$,

which shows that a^2 is self adjoint and negative (i.e., $\gamma(a^2v,v) = -\gamma(av,av) \leq 0$, equality holding for $v = 0$ only).

By elementary linear algebra, a^2 has only real negative (not necessarily different) proper values $-\lambda_1^2,\ldots,-\lambda_{2n}^2$, $\lambda_\alpha > 0$ $(\alpha=1,\ldots,2n)$, and is diagonalizable, i.e., S has a γ-orthonormal basis of proper vectors $u_1,\ldots,u_{2n}$ of a^2. This allows us to define an automorphism $p : S \to S$, whose matrix with respect to the basis u_α is diagonal:

$$\begin{pmatrix} \lambda_1 & \cdots & 0 \\ \vdots & \ddots & \vdots \\ 0 & \cdots & \lambda_{2n} \end{pmatrix} .$$

p is the unique *positive* selfadjoint operator such that $p^2 = -a^2$ (i.e., $\gamma(pv,v) \geq 0$ with equality for $v = 0$ only).

Now let us define

(2.3.7) $J = ap^{-1} = p^{-1}a$,

where the last equality holds since the two factors are diagonalizable with respect to the same basis u_α. From (2.3.7) and $p^2 = -a^2$, we get $J^2 = -\mathrm{Id}$. Furthermore, we have

$$\omega(Jv,Jw) = \omega(ap^{-1}v,ap^{-1}w) = -\omega(p^{-1}v,a^2p^{-1}w) =$$

$$= \omega(p^{-1}v,pw) = \omega(pp^{-1}v,w) = \omega(v,w) ,$$

which is the ω-compatibility of J. Finally, the metric (2.3.3) of J satisfies $g(v,v) = \omega(v,Jv) = \gamma(pv,v)$, and it is positive definite.

We might also denote J of (2.3.7) by J_γ since it is determined uniquely by γ. Notice that any positive compatible complex structure J is determined by some Euclidean scalar product on S. Indeed, we have for instance $J = J_g$ with g of (2.3.3).

Now, let us consider the structures J_0, J_1 of our theorem. Assume $J_0 = J_{\gamma_0}$, $J = J_{\gamma_1}$, and take the family of Euclidean scalar products of S, $\gamma_t = t\gamma_1 + (1-t)\gamma_0$ $(0 \leqslant t \leqslant 1)$. Then $J_t = J_{\gamma_t}$ is a differentiable family of positive, ω-compatible complex structures of S such that J_0, J_1 are the given structures.

This ends the proof of Theorem 2.3.3. Q.e.d.

In the sequel, we always refer only to positive compatible complex structures.

Remark 2.3.4. *The reason to call* g *of (2.3.3) a hermitian metric is the following one. Let us extend* g *to a hermitian scalar product* g_c *of* S_c, *in the usual sense of linear algebra, by*

(2.3.8) $$g_c(\sqrt{-1}\,u,v) = -g_c(u,\sqrt{-1}\,v) = \sqrt{-1}\,g(u,v) \qquad (u,v \in S)\ .$$

Then, the restriction of g_c *to* S' *of (2.3.2) is*

(2.3.9) $$g_c(a - \sqrt{-1}\,Ja,\ b - \sqrt{-1}\,Jb) = 2\{g(a,b) - \sqrt{-1}\,\omega(a,b)\}\ ,$$

and we see that

(2.3.10) $$h(a,b) = g(a,b) - \sqrt{-1}\,\omega(a,b) \qquad (a,b \in S)$$

is a usual hermitian metric of the complex vector space (S,J). g *appears as the real part, and* ω *as the imaginary part of* h.

Proposition 2.3.5. *The mapping*

(2.3.11) $$a \in S \mapsto \frac{1}{\sqrt{2}}(a - \sqrt{-1}\,Ja)$$

yields an isomorphism of the hermitian linear spaces (S,J,h) *and* (S',g_c) .

Proof. This follows directly from (2.3.9). Q.e.d.

Generally, we shall look at (S,J,h) and (S',g_c) as the real and complex form of the same space, the identification of vectors being given by (2.3.11).

By classical linear algebra, (S,J,h) has *orthonormal bases*, i.e., complex bases (e_i) $(i = 1,\dots,n)$ such that $h(e_i,e_j) = \delta_{ij}$. By (2.3.10) this is equivalent to

$$g(e_i,e_j) = g(Je_i,Je_j) = \delta_{ij} \ , \ g(e_i,Je_j) = 0 \ , \tag{2.3.12}$$

and we see that (e_i,Je_i) is a real g-orthonormal basis of S. Moreover, by (2.3.3), we see that $(e_i,e_{i*} = Je_i)$ is also a symplectic basis. Conversely, if (e_i,Je_i) is both a g-orthonormal and symplectic basis then e_i is a h-orthonormal basis. By (2.3.11) it has a corresponding complex basis in S' given by

$$\varepsilon_i = (e_i - \sqrt{-1}\,Je_i)/\sqrt{2} \ . \tag{2.3.13}$$

Definition 2.3.6. *A basis* (e_i,Je_i) *which is both g-orthonormal and symplectic is called a* ***real unitary basis*** *of* (S,ω,J). *The corresponding basis (2.3.13) is a* ***complex unitary basis.***

The space (S,ω,J) also has its specific groups of automorphisms which consists of linear transformations that preserve both the symplectic form ω, and the metric g (or, equivalently, commute with J). Such automorphisms are also characterized by the fact that they send a unitary basis to a unitary basis, and they are called *unitary transformations*. The group of unitary transformations is called the *unitary group* $U(S,J)$. Equivalently, the unitary group can also be seen as the group of the complex linear transformations of (S,J) that preserve the metric h of (2.3.10), or, in view of Proposition 2.3.5., as the group of complex linear transformations of S' that preserve the metric g_c.

Now, a choice of a unitary basis allows us to identify (S,J,h), and (S',g_c) with the space $\mathbb{C}^n$, via the complex coordinates (z^i) $(i = 1,\ldots,n)$ with respect to the chosen basis, while, by (2.3.12), h becomes the canonical hermitian metric $\Sigma_{i=1}^n z^i \bar{z}^i$. Then $U(S,J)$ will be identified with the usual well known *unitary group* $U(n)$ of the unitary (n,n)-matrices, i.e., complex (n,n)-matrices P such that $P^t\bar{P} = \mathrm{Id}$. $U(n)$ is known to be a compact Lie group.

On the other hand, for the Euclidean vector space (S,g), we have the *orthogonal group* $O(S,g)$ of linear isomorphisms which preserve g. By means of a g-orthonormal basis, this group is identified with the group $O(2n)$ of the orthogonal $(2n,2n)$-matrices.

Theorem 2.3.7. *Let* (S,ω) *be a real symplectic space, and* J *be a positive compatible complex structure. Then one has*

$$(2.3.14) \qquad U(S,J) = Sp(S) \cap O(S,g) .$$

Particularly, for $\mathbb{R}^{2n}$ *with the coordinates* (x^i, x^{i*}), *if* ω *is (2.1.4),* $g = \Sigma_{i=1}^n [(x^i)^2 + (x^{i*})^2]$, *and the complex structure is defined by the complex coordinates* $z^i = x^i + \sqrt{-1}\, x^{i*}$, *one has*

$$(2.3.15) \qquad U(n) = Sp(n,\mathbb{R}) \cap O(2n) .$$

$U(n)$ *is a maximal compact Lie subgroup of* $Sp(n,\mathbb{R})$.

Proof. Everything follows straightforwardly from definitions. The last assertion is a consequence of the fact that $O(n)$ is a maximal compact subgroup of $G\ell(n,\mathbb{R})$ [25]. Q.e.d.

Furthermore, let us consider again the symplectic space (S,ω), and a positive compatible complex structure J, and let $L \in \mathcal{L}(S)$ be a Lagrangian subspace of S. Then, it follows from (2.3.10), (2.3.12) that for any g-orthonormal basis (e_i) of the subspace L, (e_i, Je_i) is a real unitary basis of S.

Definition 2.3.8. *A real or complex unitary basis of* S *such that its real vectors* e_i $(i = 1,\ldots,n)$ *belong to* L *is called an* L-*related unitary basis.*

Clearly, L-related bases always exist. If also $L' \in \mathcal{L}(S)$, and (e'_i, Je'_i) is an L'-related basis, there exists a unitary transformation that sends (e_i, Je_i) to (e'_i, Je'_i), i.e., that sends L to L'. Hence, the unitary group $U(S,J)$ acts transitively on $\mathcal{L}(S)$. The corresponding isotropy subgroup of L is the group of transformations between the L-related unitary bases, and it is obviously isomorphic to the orthogonal group of (L,g). The following fundamental result is proven thereby

Theorem 2.3.9. $\mathcal{L}(S) \approx U(S,J)/O(L) \approx U(n)/O(n)$.

Here, $O(L)$ denotes the orthogonal group of (L,g), the first equivalence describes $\mathcal{L}(S)$ as a homogeneous space, and the second is determined by the choice of an L-related unitary basis.

Another interesting utilization of the compatible complex structures is in the expression of the transversality of two Lagrangian subspaces [74]. Let us identify (S,ω,J) with $\mathbb{C}^n$ by choosing a unitary basis, and let us use for the coordinates of $\mathbb{C}^n$ the same notation as in Theorem 2.3.7. Then, the *real part* $\mathbb{R}^n = \{(x^i)\}$ of $\mathbb{C}^n$ is a Lagrangian subspace, and, by Theorem 2.3.9, every Lagrangian subspace X of $\mathbb{C}^n$ (which corresponds to a Lagrangian subspace of S) can be represented as $X = A(\mathbb{R}^n)$ with $A \in U(n)$. Furthermore, $A(\mathbb{R}^n) = B(\mathbb{R}^n)$ $(A,B \in U(n))$ iff $B^{-1}A$ is an orthogonal matrix (by Theorem 2.3.9), and, since this is a unitary matrix, it is orthogonal iff $B^{-1}A$ is a real matrix or equivalently

(2.3.16) $\quad A\bar{A}^{-1} = B\bar{B}^{-1}$.

Accordingly, we have a well defined *Souriau mapping* $\sigma: \mathcal{L}(S) \to U(n)$ given by $\sigma(X) = A\bar{A}^{-1}$. (It depends, however, on the choice of the unitary basis in S.)

Proposition 2.3.10. *Two Lagrangian subspaces* $X, Y \in \mathcal{L}(S)$ *are transversal iff the matrix* $\sigma(X) - \sigma(Y)$ *is a nondegenerate matrix.*

Proof. We may work in $\mathbb{C}^n$ with $X = A(\mathbb{R}^n)$, $Y = B(\mathbb{R}^n)$. Then $x \in X$ iff $x = Ar$ for a real vector $r \in \mathbb{R}^n$, which is equivalent to $x = \sigma(X)\bar{x}$. Similarly, $y \in Y$ iff $y = \sigma(Y)\bar{y}$. Hence $X \cap Y \neq \{0\}$ iff there is a common solution $z \neq 0$ for $z = \sigma(X)\bar{z}$, $z = \sigma(Y)\bar{z}$, which implies that $\sigma(X) - \sigma(Y)$ is a degenerate matrix. Conversely, let this happen, and take $A = Id.$ (which is enough). Then, for some $z \neq 0$, $\bar{z} = B\bar{B}^{-1}\bar{z}$, whence $z = B\bar{B}^{-1}z$, and either Re z or Im z or both are nonzero and in $X \cap Y$. Q.e.d.

Corollary 2.3.11. *Let* $L_1, \ldots, L_k$ *be a finite number of Lagrangian subspaces of a real symplectic vector space. Then there exists a Lagrangian subspace transversal to all the subspaces* $L_1, \ldots, L_k$.

Proof. It is clearly possible to choose a matrix A such that $\sigma(L) - \sigma(L_a)$ be nondegenerate for all a=1,...,k, where L is defined by $A(\mathbb{R}^n)$. Q.e.d.

Now, we shall finish by a discussion of the set of all the positive compatible complex structures J of a real symplectic space (S,ω). We shall denote this set by $\mathcal{J} = \mathcal{J}(S,\omega)$. It can be described in various ways [25], [37], and we choose to give here only its description by means of the so-called *Siegel upper half-plane* [72], [25].

For a fixed $J \in \mathcal{J}$, (S,J) is identifiable by Proposition 2.3.5 with a subspace $S' \subset S_c$. It is important to notice that, if we consider the complex symplectic space (S_c, ω) where ω is extended to

S_c by linearity, then S' is a Lagrangian subspace. Indeed, because of (2.3.1), we have

$$\omega(a - \sqrt{-1}\, Ja,\ b - \sqrt{-1}\, Jb) = \omega(a,b) - \omega(Ja,Jb) - $$
$$- \sqrt{-1}\big(\omega(Ja,b) + \omega(a,Jb)\big) = 0 \ .$$

Another important property follows from (2.3.3):

(2.3.17) $$-\sqrt{-1}\ \omega(v,\bar{v}) = g(a,a) > 0$$

for every vector $0 \neq v = (a - \sqrt{-1}\ Ja)/\sqrt{2} \in S'$. Finally, we have from (2.3.2) that $S_c = S' \oplus \bar{S}'$, and if $v \in S_c$ is $v = v_1 + v_2$, $v_1 \in S'$, $v_2 \in \bar{S}'$ we get

(2.3.18) $$Jv = \sqrt{-1}\ (v_1 - v_2) \ .$$

This suggests

Definition 2.3.12. *A Lagrangian subspace* F *of the complex symplectic space* (S_c,ω) *is called* **positive** *if* $-\sqrt{-1}\ \omega(v,\bar{v}) > 0$ *for every* $0 \neq v \in F$.

Proposition 2.3.13. *There exists a natural bijection between the set* $J(S,\omega)$, *and the set* $F_+(S_c)$ *of all the positive Lagrangian subspaces of* (S_c,ω).

Proof. The mapping $J \to F_+$ is defined by $J \mapsto S'$. If $F \in F_+$, then, since ω is a real form, $\bar{F}$ is also a Lagrangian subspace of S_c. Moreover, it follows easily from the positivity of F that $F \cap \bar{F} = \{0\}$, which implies $S_c = F \oplus \bar{F}$. Now, we may define $J : S_c \to S_c$ by

(2.3.19) $$J(f_1 + \bar{f}_2) = \sqrt{-1}(f_1 - \bar{f}_2) \qquad (f_1, f_2 \in F) \ ,$$

and it satisfies $J^2 = -\mathrm{Id.}$, $\omega(Jv,Jw) = \omega(v,w)$ for all $v,w \in S_c$. Since

$v \in S_c$ satisfies $v \in S$ iff $v = \bar{v}$, we also see that $J(S) = S$. Finally, for $0 \neq v = f_1 + \bar{f}_1 \in S$ (necessarily $f_2 = f_1$), we have $\omega(v,Jv) = -2\sqrt{-1}\omega(f_1,f_1) > 0$. Hence J is a positive compatible complex structure on (S,ω), and we obtain thereby the mapping $F_+ \to J$. From (2.3.18) and (2.3.19) it follows easily that the two mappings $J \to F_+$, $F_+ \to J$ defined above are inverse to each other. Q.e.d.

Now, in order to go on with the description of J let us make the following considerations. A Lagrangian subspace L_c of the complex symplectic vector space S_c above is called a *real Lagrangian subspace* if it is the complexification of a Lagrangian subspace $L \subset S$, i.e., $L_c = L \otimes_{\mathbb{R}} \mathbb{C}$. It is easy to see that this happens iff the complex conjugate subspace is $\bar{L}_c = L_c$. In S_c, any real Lagrangian subspace L_c is transversal to any positive Lagrangian subspace $F \in F_+$. Indeed, if

$$0 \neq v = a + \sqrt{-1}\, b \in L_c \cap F \qquad (a,b \in L) ,$$

then

$$0 < -\sqrt{-1}\,\omega(v,\bar{v}) = -2\omega(a,b) = 0 ,$$

which is a contradiction.

Hence, if we fix L_c, $F_+ \subset T(L_c)$, and if L'_c is another real Lagrangian subspace of S_c which is transversal to L_c in S_c (i.e., $L \oplus L' = S$), we may apply the results encountered during the proof of Theorem 2.2.15. Namely, the pair L'_c, $F \in F_+$ defines a vector of the affine space $T(L_c)$, i.e., a quadratic form $Q(L_c,F)$ on L^*_c associated to the unique symplectic transformation σ that fixes every point of L_c, and sends L'_c onto F. If (e_i,e_{i*}) $(e_i \in L,\ e_{i*} \in L')$ is a fixed symplectic basis adapted to L, $Q(L_c,F)$ is represented by the (n,n)-complex symmetric matrix $Z = M$ defined by the matrix of σ in the form (2.3.13). Moreover, this matrix Z is nondegenerate since F is transversal to L'_c as well. Finally, we must ask Z to be such as to ensure the positivity of F. In view of (2.2.13), the transformation σ is defined by

$$(2.3.20) \qquad \sigma(e_i) = e_i, \ \sigma(e_{i*}) = e_j z^j_{i*} + e_{i*} \ ,$$

where (z^j_{i*}) are the components of the symmetric matrix Z. $\sigma(e_{i*})$ is a basis in F, and a simple computation shows that $-\sqrt{-1}\,\omega(v,\bar{v}) > 0$ for $0 \neq v \in F$ iff the imaginary part of the matrix Z, i.e.,

$$(2.3.21) \qquad \mathrm{Im}\, Z = \frac{1}{2\sqrt{-1}}(Z - \bar{Z})$$

is a positive definite matrix $(\mathrm{Im}\, Z > 0)$.

The following notion is well known in complex function theory [72].

Definition 2.3.14. *The space* S_n *of all the symmetric complex* (n,n)*-matrices* Z *such that* $\mathrm{Im}\, Z > 0$ *is called the* n*-th* ***Siegel upper half-plane.***

Therefore, above we established a mapping $F_+ \to S_n$ given by $F \mapsto Z$, and defined by the choice of a symplectic basis (e_i, e_{i*}) in S. It is clear that, once this basis is choosen, the mapping obtained is a bijection since Z determines the symplectic transformation σ of (2.3.20). Moreover, if the symplectic basis is changed by

$$(2.3.22) \qquad (e\ e_*) = (\tilde{e}\ \tilde{e}_*) \begin{pmatrix} A & B \\ C & D \end{pmatrix} ,$$

then, we can easily compute the relation between the two matrices Z, $\tilde{Z}$ associated to a same space $F \in F_+$. Namely, by (2.3.20) F has the bases $eZ + e_*$ and $\tilde{e}\tilde{Z} + \tilde{e}_*$, hence

$$(2.3.23) \qquad eZ + e_* = (\tilde{e}\tilde{Z} + \tilde{e}_*)\Lambda$$

for some nondegenerate (n,n)-matrix Λ. If in (2.3.23) we insert e, e_*

as given by (2.3.22), we get $\Lambda = CZ + D$, and then

$$\tilde{Z} = (AZ + B)(CZ + D)^{-1} , \tag{2.3.24}$$

which is a *complex analytic automorphism* of S_n [72].

Now, if we combine the bijection $F_+ \to S_n$ with the result of Proposition 2.3.13, we get

Theorem 2.3.15. *For a real symplectic space* (S^{2n},ω), *there is an identification of the set* $J(S,\omega)$ *of the positive compatible complex structures with the Siegel upper half-plane* S_n, *which is defined up to an analytical automorphism of* S_n.

Remark 2.3.16. *The identification of Theorem 2.3.15 can also be seen as a topological identification.*

Finally, let us give one more interesting consequence of Proposition 2.3.13. The action of any transformation $\sigma \in Sp(S)$ extends naturally to S_c, and using Definition 2.3.12 we see that σ also acts on F_+. This defines a transitive action of $Sp(S)$ on F_+. Indeed, the form

$$h(v,w) = -\sqrt{-1}\,\omega(v,\bar{w}) \qquad (v,w \in F \in F_+) \tag{2.3.25}$$

is a hermitian metric on F, and we may distinguish h-orthonormal bases f_i in F. Then $(f_i,-\sqrt{-1}\,\bar{f}_i)$ is a symplectic basis in S_c. Similarly, for $F' \in F_+$ we shall have a symplectic basis $(f'_i,-\sqrt{-1}\,f'_i)$ with $f'_i \in F'$. The transformation $\sigma(f_i) = f'_i$, $\sigma(\bar{f}_i) = \bar{f}'_i$ is then symplectic in S_c, it sends F to F', and it commutes with complex conjugation. Therefore, $\sigma \in Sp(S)$, which proves the transitivity property mentioned.

Furthermore, it is easy to understand that the isotropy subgroup of F is the unitary group $U(F,h)$ with h given by (2.3.25). Hence, we get

$$J(S,\omega) \approx F_+(S\ ) \approx Sp(S)/U(F,h) \approx Sp(n,\mathbb{R})/U(n) \quad , \tag{2.3.26}$$

where the last equivalence follows by a choice of bases.

Formula (2.3.26) also shows that $J(S,\omega)$ is diffeomorphic to a cell [37].

CHAPTER 3

SYMPLECTIC GEOMETRY ON MANIFOLDS

We understand the symplectic geometry on manifolds not just as the geometry of the symplectic manifolds but as the study of various geometric structures on manifolds where a symplectic vector bundle plays an essential role. In the present chapter, we shall discuss about such structures. Of course, symplectic manifolds are the core of this theory. We shall also discuss the Lagrangian subbundles naturally related with the considered structures.

3.1. SYMPLECTIC VECTOR BUNDLES. The notion of a symplectic vector bundle has been defined in Definition 1.1.3. A symplectic vector bundle is a vector bundle $\pi : E \to M$, where the basis M is an m-dimensional differentiable manifold, the fiber of E is $\mathbb{R}^{2n}$, and the bundle is endowed with a differentiable field ω of symplectic forms of the fibers of E. Equivalently, ω is a cross section of $\Lambda^2 E^*$ whose restrictions to the fibers of E are nondegenerate. Another equivalent definition is that a symplectic structure on $\pi : E \to M$ is a reduction of the structure group to $sp(n, \mathbb{R})$. Consequently, by Proposition 2.1.8a), a symplectic vector bundle is orientable.

The geometry of the symplectic vector spaces as described in Chapter 2 extends naturally, fiberwise, to symplectic vector bundles. In every fiber, we have the set of the corresponding symplectic bases, and the union of these sets over M yields a principal fibre bundle, $S(E)$ with the structure group $sp(n, \mathbb{R})$. The notion of skeworthogonality (Definition 2.1.5), and the isomorphism $\flat$ of (2.1.10) will now be defined fiberwise. If F is a vector subbundle of E of constant fiberwise symplectic rank we can use, fiberwise, the definitions (2.2.1), (2.2.2), (2.2.4), and get the *skeworthogonal subbundle* ${}^{s}F$, the *radical subbundle* ${}^{i}F = F \cap {}^{s}F$, and the *union subbundle* ${}^{u}F = F + {}^{s}F$. Because of the continuity of ω, the reduction operation is meaningful for vector bundles as well, which, by (2.2.5) yields the *reduced symplectic bundle* ${}^{r}F = F/{}^{i}F$.

Furthermore, always applying the definitions fiberwise, we shall distinguish isotropic, coisotropic, Lagrangian, and symplectic subbundles. Then, it is obvious that Proposition 2.2.11 holds if, everywhere in its formulation, vector spaces are replaced by vector bundles over M. It is also very important to mention that the vector bundle variant of Theorem 2.2.4 holds good. The only delicate point is the global definition of the subbundle Q there, but this is settled by Remark 2.2.5.

Let $(\pi : E \to M, \omega)$ be a symplectic vector bundle, and let us denote by $\mathcal{L}(E_x)$ the set of the Lagrangian subspaces of the fibre E_x $(x \in M)$. Then

$$\pi' : \mathcal{L}(E) = \bigcup_{x \in M} \mathcal{L}(E_x) \to M , \tag{3.1.1}$$

with the natural projection π', is again a locally trivial bundle, and we call it the *Lagrange Grassmannian bundle of* E. Indeed, in view of Proposition 2.2.13, and of its proof, $\mathcal{L}(E)$ is the bundle which enters in the following well known diagram of bundles (e.g., [80]).

$$\begin{array}{ccc} S(E) & \overset{\sigma}{\to} & S(E)/G_{L_0} = \mathcal{L}(E) \\ & {}_{\pi}\searrow \quad \swarrow_{\pi'} & \\ & M & \end{array} \tag{3.1.2}$$

where G_{L_0} is the group of the symplectic matrices (2.2.10), and it acts on $S(E)$ at the right.

A Lagrangian subbundle L of E is equivalent with a global cross-section $L : M \to \mathcal{L}(E)$. Consequently, such a subbundle exists iff the structure group of $S(E)$ can be reduced to G_{L_0} [80]. This does not always happen. For instance, let S^2 be the unit sphere, and ω be its usual volume form. Then (TS^2, ω) is a symplectic vector bundle over S^2, with 2-dimensional fiber, and any subbundle of rank 1 is a Lagrangian subbundle. But it is well known that such a subbundle does not exist since S^2 has a nonzero Euler-Poincaré characteristic.

If the bundle E has a Lagrangian subbundle L, then, by using the vector bundle variant of Theorem 2.2.4 we will be able to construct a *transversal Lagrangian subbundle* L' (transversality is again defined fiberwise) such that $E = L \oplus L'$. A choice of (L,L') reduces the structure group of E to the group of the symplectic matrices 2.2.12, hence we have

Proposition 3.1.1. *The symplectic vector bundle* E *has a Lagrangian subbundle iff its structure group can be reduced to* $G\ell(n,\mathbb{R})$ *(realized by matrices (2.2.12)).*

Remarks 3.1.2. *1) The above decomposition* $E = L \oplus L'$ *yields an isomorphism* $E \simeq L \oplus L^*$ *similar to the one encountered for vector spaces.* *2) The various interesting bases encountered for symplectic vector spaces can be extended to bases of local cross sections in the case of vector bundles.*

Furthermore, if E has a Lagrangian subbundle L, we have for every $x \in M$ the affine space $T(L_x)$ of the transversal Lagrangian subspaces to L_x in E_x, and we may define

$$T(L) = \bigcup_{x \in M} T(L_x) . \qquad (3.1.3)$$

Looking at the proof of Theorem 2.2.15, we see immediately that $\mathcal{T}(L)$ is an affine bundle over M whose fiber is the real $n(n+1)/2$-dimensional affine space. We call $\mathcal{T}(L)$ the *affine transversal bundle of* L.

Now, let us generalize the results of Section 2.3. The definitions of compatible (positive) complex structures, and hermitian metrics carry over to vector bundles fiberwise. Of course instead of single tensors, we shall have a field $J(x)$ which satisfies (2.3.1) for all $x \in M$, and a field $g(x)$ of hermitian metrics given by (2.3.3). Theorem 2.3.3 is generalized by

Theorem 3.1.2. *Every symplectic vector bundle has compatible positive complex structures* J, *and hermitian metrics* g, *and any two such structures can be joined by a differentiable curve* J_t, $0 \leq t \leq 1$.

The proof is the same as for Theorem 2.3.3 if we use a global Euclidean scalar product γ on E, which is known to exist always.

It is also easy to obtain the bundle variant of Proposition 2.3.5, and to define, like in Definition 2.3.6, *real and complex unitary bases of local sections of a bundle* (E,J), which will be related by (2.3.13). It is clear from these definitions that if, for $x \in M$, we denote by $\mathcal{U}(E_x,J_x)$ the set of (real or complex) unitary bases of the fibre (E_x,J_x,g_x), and if

$$\mathcal{U}(E,J) = \bigcup_{x\in M} \mathcal{U}(E_x,J_x), \tag{3.1.4}$$

then $\mathcal{U}(E,J)$ has a natural structure of a principal bundle with structure group the unitary group $U(n)$. $\mathcal{U}(E,J)$ is a subbundle of $\mathcal{S}(E)$, hence it follows

Proposition 3.1.3. *A choice of a compatible positive complex structure on* E *is equivalent with a reduction of the structure group of* E *to the unitary group* U(n).

Remark 3.1.4. *The existence of such a reduction follows also from general theorems about fiber bundles since* U(n) *is a maximal compact subgroup of* $Sp(n,\mathbb{R})$ *(Theorem 2.3.7). This is another proof of Theorem 3.1.2.*

In the sequel, we agree that all the compatible complex structures which we shall use are positive. We shall also use the term of L-*related unitary bases* in the sense of Definition 2.3.8 for Lagrangian subbundles L.

In agreement with Theorem 2.3.9, and by identifying a Lagrangian subspace with the set of its related unitary bases, we have now

Proposition 3.1.5. *The Lagrange-Grassmannian bundle* $\mathcal{L}(E)$ *of the symplectic vector bundle* E *is also associated to the principal bundle* $U(E,J)$, *by the commutative diagram of bundles.*

$$(3.1.5)\qquad U(E,J) \xrightarrow{\sigma} U(E,J)/_{O(n)} = \mathcal{L}(E), \qquad \pi : U(E,J) \to M, \quad \pi' : \mathcal{L}(E) \to M$$

The fibre of $\mathcal{L}(E)$ *is* U(n)/O(n), *and the structure group of* $\mathcal{L}(E)$ *is reduced to* U(n).

Now, it follows from (3.1.5), and general fiber bundle theory

Proposition 3.1.6. E *admits a Lagrangian subbundle (i.e., a global cross cross-section* $L : M \to \mathcal{L}(E)$ *if the structure group of* E *can be reduced to the orthogonal group* O(n).

This result can also be understood as follows. We start with the principal bundle $\mathcal{S}(E)$ and construct its subbundle $\mathcal{U}(E,J)$ after a choice of J has been made. Then, if L is a Lagrangian subbundle of E, we look at the set $\mathcal{U}(E_x,J_x;L_x)$ of all the L_x-related unitary bases $(x \in M)$, and at

$$\mathcal{U}(E,J;L) = \bigcup_{x\in M} \mathcal{U}(E_x,J_x;L_x) \quad . \tag{3.1.6}$$

By looking at the proof of Theorem 2.3.9, it follows that $\mathcal{U}(E,J;L)$ has a natural structure of an $O(n)$-principal subbundle of $\mathcal{U}(E,J)$ over M, which is precisely the reduction of Proposition 3.1.6. Notice that $O(n)$ is seen here as the group of the matrices (2.2.12) where $A \in O(n)$, and therefore ${}^tA^{-1} = A$. It also follows from this construction that

$$E \approx L \oplus JL \; . \tag{3.1.7}$$

Finally, if we are interested in all the compatible positive complex structures of E, we have an obvious (pointwise) generalization of Definition 2.3.12 which gives us the notion of a positive Lagrangian subbundle F of the complexified bundle E_c. Furthermore, the bundle version of Proposition 2.3.13 obviously holds as well, which means that there is a bijection between the set $\mathcal{J}(E,\omega)$ of compatible positive complex structures of the symplectic vector bundle (E,ω) and the set $\mathcal{F}_+(E_c)$ of the positive Lagrangian subbundles of (E_c,ω).

On the other hand, Theorem 2.3.15 holds on every fiber of E, and, in fact, it holds for sections over trivializing neighbourhoods of E. It follows that the set

$$\Sigma(E) = \bigcup_{x\in M} \mathcal{J}(E_x,\omega_x) \tag{3.1.8}$$

is a locally trivial fiber bundle over M whose fibre is the Siegel half plane S_n, and whose structure group is the group of the analytical automorphisms (2.3.28) of S_n.

Definition 3.1.7. $\Sigma(E)$ *is called the* ***Siegel bundle*** *of the symplectic vector bundle* E.

Of course, a compatible positive complex structure of E is equivalent to a global cross section of the Siegel bundle $\Sigma(E)$.

3.2. SYMPLECTIC VECTOR BUNDLES OF GEOMETRIC STRUCTURES. In this Section, we discuss the main goemetric structures based on symplectic vector bundles.

Definition 3.2.1. *Let* V *be a differentiable manifold. A symplectic structure of the tangent bundle* TV *is called an* ***almost symplectic structure*** *of* V*, and* V *endowed with this structure is an* ***almost symplectic manifold.***

Clearly, an almost symplectic structure can exist only if $\dim V = 2n$ is even, and the structure is equivalent with a global nowhere degenerate 2-form ω on V. Almost symplectic structures may not exist on some even dimensional manifolds. We know from Section 3.1 that a necessary condition for almost symplectic manifolds is orientability. From Theorem 3.1.2, we see that V has an almost symplectic structure iff it has an almost complex structure. Since such structures exist only on S^2 and S^6 [45], it follows that the unit spheres S^m with $m \neq 2,6$ do not possess almost symplectic structures. For the geometric study, and classification of the almost symplectic manifolds, we refer the reader to [52], [33].

As we saw in Chapter 1, the important case for applications is that of the *symplectic manifolds* (V,ω) where the 2-form ω is not only nondegenerate but also closed (Definition 1.1.3). We saw in Chapter 1 that the basic example of a symplectic manifold is a contanget bundle T^*M with the canonical symplectic form Ω of (1.1.5). Particularly, we have here the symplectic manifold $\mathbb{R}^{2n} = T^*\mathbb{R}^n$. Related to this example it is interesting to notice that T^*S^{n-1} is diffeomorphic to the manifold $\Delta(\mathbb{R}^n)$ of all the oriented straight lines of $\mathbb{R}^n$. Indeed, such a line ℓ is determined by its unit vector, i.e., a point $x_0 \in S^{n-1}$, and by the intersection point of radius vector r_0 of ℓ with the tangent plane of S^{n-1} at x_0. It follows that $\Delta(\mathbb{R}^n)$ has a natural structure of a $2(n-1)$-dimensional symplectic manifold. Namely, if we identify the tangent and cotangent planes of S^{n-1} by means of the metric induced by $\mathbb{R}^n$, the Liouville form $\lambda(S^{n-1})$ (see (1.1.4)) is equal to the scalar product $(r_0-x_0)dx_0 = r_0 dx_0$ in $\mathbb{R}^n$ ($x_0^2 = 1$ implies $x_0 dx_0 = 0$). Then, the symplectic form (1.1.5) is $\Omega(S^{n-1}) = dx_0 \wedge dr_0$ seen as an exterior scalar product of vector forms. This is precisely the symplectic form of $\Delta(\mathbb{R}^n)$. Moreover, if r is the radius vector of an arbitrary point of ℓ, we have $r = r_0 + \alpha x_0$, and using again $x_0 dx_0 = 0$, we also get $\lambda(S^{n-1}) = rdx_0$, and $\Omega = dx_0 \wedge dr$.

Many other interesting symplectic manifolds also exist.

For instance, let us consider a dynamical system with phase space T^*M, and the Hamiltonian function H (Section 1.3). Then, the orbits of the system are those of the vector field X_H defined by (1.3.7), and, particularly, the orbits sitting on a hypersurface of constant energy $H = \text{const.}$ satisfy $i(X_H)\Omega = 0$ on the hypersurface. Remember that we also have $L_{X_H}\Omega = 0$ (Proposition 1.2.7). Generally, the orbits of X_H on $H = \text{const.}$ define a foliation (with singularities), but it may happen that these orbits be the fibers of a locally trivial fibration of $H = \text{const.}$ over a certain basis B. Then, $i(X_H)\Omega = L_{X_H}\Omega = 0$ prove that Ω/H induces a symplectic form on B, and B becomes a symplectic manifold.

This example can be generalized in a construction of symplectic manifolds by *reduction* [87], [1]. As in Section 1.2, we have

Definition 3.2.2. *Let* N *be a* (2n+k)-*differentiable manifold* ($k \geq 1$), *endowed with a closed 2-form* $\hat{\omega}$ *which is everywhere of* rank 2n. *Then* $\hat{\omega}$ *is called a* ***presymplectic form***, *and* $(N,\hat{\omega})$ *is a* ***presymplectic manifold.***

Let $(N,\hat{\omega})$ be a presymplectic manifold. As in Example c) of Section 2.1, $\forall x \in N$, we have the annihilator space

$$(3.2.1) \qquad F_x = \{v \in T_xN / i(v)\hat{\omega}_x = 0\} \quad ,$$

and they generate a k-dimensional distribution (subbundle of TN) F on N, called the *annihilator of* $\hat{\omega}$.

Proposition 3.2.3. *The distribution* F *defined by (3.2.1) is integrable, and* $\hat{\omega}$ *is a foliate form with respect to the foliation defined by* F.

Proof. Let X,Y be vector fields on N belonging to F, i.e.,

$$(3.2.2) \qquad i(X)\hat{\omega} = 0, \quad i(Y)\hat{\omega} = 0 \ .$$

Then we have [59]

$$(3.2.3) \qquad i([X,Y])\hat{\omega} = L_X i(Y)\hat{\omega} - i(Y)L_X\hat{\omega} =$$

$$= 0 - i(Y)\left(i(X)d\hat{\omega} + di(X)\hat{\omega}\right) = 0 \quad ,$$

which proves the integrability of F. Hence F defines a *foliation* of N whose *leaves* are the maximal connected integral manifolds of F given by the global Frobenius theorem [59].

On a *foliated manifold* (i.e., a manifold endowed with a foliation, the functions, forms, etc. that depend only on the leaves are labelled as *foliate* [80] (or *projectable* or *base-like*). If F is the foliation, then the Frobenius theorem shows that every point $x \in N$ has a cubical neighbourhood centered at x, and endowed with local coordinates

(y^a, x^u) $(a,b,\ldots = 1,\ldots,k;\ u,v,\ldots = 1,\ldots,2n)$ such that the leaves of F are locally defined by $x^u = \text{const}$. These coordinates will be called *adapted to* F, y^a will be *leaf coordinates* (locally, $F = \text{span}\,\{\partial/\partial y^a\}$), x^u are *transverse coordinates*, and the transition functions between two systems of adapted coordinates are locally of the form

$$\tilde{x}^u = \tilde{x}^u(x^v),\ \tilde{y}^a = \tilde{y}^a(x^u, y^a) \quad . \tag{3.2.4}$$

A function, form, etc. is foliate iff its local coordinate expressions depend locally on the coordinates x^u only.

In our case, in view of the definitions, $\hat{\omega}$ satisfies

$$i(X)\hat{\omega} = 0\ ,\quad L_X\hat{\omega} = di(X)\hat{\omega} + i(X)d\hat{\omega} = 0 \tag{3.2.5}$$

for every vector field X in F. By means of the adapted local coordinates, (3.2.5) is equivalent to the fact that locally

$$\hat{\omega} = \frac{1}{2}\hat{\omega}_{uv}(x^w)dx^u \wedge dx^v \quad . \tag{3.2.6}$$

Q.e.d.

Generally, the space of the leaves of a foliation is not a manifold or may be a non-Hausdorff manifold, and we shall give

Definition 3.2.4. *Let* F *be a foliation of* N, *and let us denote by* N/F *the space of its leaves. If* N/F *is a Hausdorff differentiable manifold, and the natural projection* $\pi\colon N \to N/F$ *is a submersion, we shall say that* F *is a* ***nice foliation***. N/F *is also called the* ***base manifold*** *of* (N,F).

Now, we have the following immediate consequence of Proposition 3.2.3.

Corollary 3.2.5. *If* $(N,\hat{\omega})$ *is a presymplectic manifold, and if the annihilator* F *of* $\hat{\omega}$ *is a nice foliation of* N, *then* $\hat{\omega}$ *projects to a symplectic structure* ω *on the base manifold* N/F.

Definition 3.2.6. $(N/F,\omega)$ *of Corollary 3.2.5. is called the* ***symplectic reduction*** *of* $(N,\hat{\omega})$.

Symplectic reduction plays an important role in the study of symplectic manifolds endowed with a differentiable action of a Lie group G by *symplectomorphisms*, i.e., diffeomorphisms that preserve the symplectic structure. The physical applications are to mechanical systems with symmetry [1].

The previous remark suggests the interesting problem of the determination of the *homogeneous symplectic manifolds*, i.e., symplectic manifolds admitting a transitive Lie group of symplectomorphisms. The interested reader can find a discussion of this subject in [37]. Here, we would like to give only a short description of the main example.

Let G be a connected Lie group with the Lie algebra g and its dual space g^*. If we set

$$\langle\gamma(a),X\rangle = \langle a,(\mathrm{ad}\gamma^{-1})X\rangle \qquad (a \in g^*,\ X \in g,\ \gamma \in G) \tag{3.2.7}$$

we obtain an action of G on g^* called the *coadjoint action*, and whose orbits are called *coadjoint orbits*. Similarly, there is a coadjoint action of g on g^* which is defined by

$$\langle Y(a),X\rangle = \langle a,[Y,X]\rangle \quad , \tag{3.2.8}$$

where $X,Y \in g$ and the bracket is the one of the Lie algebra g.

Now, let us fix $a \in g^*$, and let $\mathcal{O}_a = \mathcal{O}$ be its coadjoint orbit. Then G acts transitively on $\mathcal{O}$, and a has the isotropy subgroup $G_a = \{\gamma \in G/\gamma(a) = a\}$ with the corresponding Lie algebra

$g_a = \{X \in g / X(a) = 0\}$. Hence $\mathcal{O}$ can be seen as the manifold G/G_a.

Now, on G we can define the 2-form $\hat{\omega}$ to be the unique left invariant 2-form such that

(3.2.9) $$\hat{\omega}_e(X,Y) = \langle a, [X,Y] \rangle \quad ,$$

where e is the unit of G, and $X,Y \in T_e G = g$, with the bracket operation of g. This form is obviously of a constant rank, and a straightforward computation, which uses the Jacobi identity in g, shows that $(d\hat{\omega})_e = 0$, hence $d\hat{\omega} = 0$. Therefore $(G,\hat{\omega})$ is a presymplectic manifold.

Furthermore, it is clear from (3.2.9) that the annihilator of $\hat{\omega}$ is the left invariant distribution generated by $g_a \subset T_e G$. Hence, the space of the leaves is precisely G/G_a, and the considered distribution defines a nice foliation of G. Accordingly, by applying Corollary 3.2.5, we get

Theorem 3.2.7. (Kirillov, Kostant, Souriau). *Any coadjoint orbit of a connected Lie group* G *is a homogeneous symplectic manifold.*

There are also other, more exotic, examples of symplectic manifolds such as the space of binary forms of odd degree in two variables, x, y, and the space of monic polynomials of even degree in x [5].

The existence and global classification of symplectic structures on manifolds is a difficult topological problem with many still open aspects [36]. For open manifolds V^{2n}, M.Gromov [35] proved that any almost symplectic structure of V (if any) is homotopic to a symplectic structure. For compact 2-manifolds, J. Moser [67] proved that two symplectic manifolds (V_a^2, ω_a) (a=1,2) are symplectomorphic iff they have the same genus and the same volume.

Before we proceed, let us mention one more interesting class of almost symplectic manifolds. Hamilton's formulation of mechanics (Section 1.3) means that a dynamical system consists of a phase space, which is an evendimensional differentiable manifold V, and a dynamical vector field X on V, whose orbits are the motion orbits, and such that, with respect to some local canonical coordinates (q^i, p_i), these orbits are given by Hamilton's equations (1.3.4). But this viewpoint does not ask necessarily for the existence of a global symplectic form Ω equal to (1.3.5) in the above-mentioned local canonical coordinates. Indeed, if V has an atlas whose transition functions

$$\tilde{q}^i = \tilde{q}^i(q^j, p_j) \ , \ \tilde{p}_i = \tilde{p}_i(q^j, p_j)$$

imply $\tilde{\Omega} = \lambda\Omega$ (λ = const. $\neq 0$) where Ω is defined by (1.3.5), and if we consider associated local Hamiltonian functions H with transition relations $\tilde{H} = \lambda H$ (same λ as above), then it follows easily that the local symplectic gradients $X_H, X_{\tilde{H}}, \ldots$ (Definition 1.2.1), which satisfy the Hamilton equations, glue up to a global dynamical vector field on V.

Hence, a generalized mathematical model of a mechanical system may be considered as a manifold V endowed with an open covering $\{U_\alpha\}_{\alpha \in A}$, and a system of local symplectic forms Ω_α on U_α such that over $U_\alpha \cap U_\beta \neq \emptyset$ one has

(3.2.10) $$\Omega_\beta = \lambda_{\beta\alpha}\Omega_\alpha \qquad (\lambda_{\beta\alpha} = \text{const.}).$$

From (3.2.10) it follows $\lambda_{\beta\alpha}\lambda_{\alpha\gamma} = \lambda_{\beta\gamma}$, and therefore $\lambda_{\beta\alpha} = e^{\sigma_\alpha}/e^{\sigma_\beta}$ for some local functions $\sigma_\alpha : U_\alpha \to \mathbb{R}$. Accordingly, V is endowed with a global nowhere degenerate 2-form Ω such that

(3.2.11) $$\Omega/U_\alpha = e^\sigma \, \Omega_\alpha \quad ,$$

or, equivalently (by differentiating (3.2.11))

(3.2.12) $$d\Omega = \theta \wedge \Omega \quad ,$$

where θ is a global closed 1-form locally equal to $d\sigma_\alpha$.

Definition 3.2.8. *An almost symplectic manifold* (V,Ω) *is called a **locally conformal symplectic manifold** if* Ω *satisfies one of the equivalent properties (3.2.11) or (3.2.12).*

For the study of such manifolds see [55], [82].

Let us go on with our discussion of structures which have symplectic vector bundles. First, let us come back to the presymplectic manifolds $(N,\hat{\omega})$. The reason is that even if the annihilator foliation F of $\hat{\omega}$ is not nice, the manifold is still equipped with an interesting symplectic vector bundle namely, $(TN/F,\hat{\omega})$. Moreover, let H be a fixed supplementary distribution of F, i.e., $TN = H \oplus F$. Then $(H,\hat{\omega})$ is again a symplectic vector bundle isomorphic to $(TN/F,\hat{\omega})$. This is particularly important if a distribution H can be defined by some geometric conditions. Then, it is usual to refer to F as the *vertical distribution* and to H as the *horizontal distribution.*

Most naturally, presymplectic manifolds appear as submanifolds (N,ω_N) of symplectic manifolds (V,ω) where ω_N is induced by ω. It suffices to ask ω_N to be of constant rank, in which case we say that N is a *submanifold of constant rank* of V. (In the sequel, we shall also use the labels *isotropic, coisotropic, Lagrangian* and *symplectic* for submanifolds N of (V,ω) whose tangent bundles are isotropic,..., etc. subbundles of TV/N.) As a matter of fact, all presymplectic manifolds appear in this way. Namely, one has [32] (see also [56], [81]).

Theorem 3.2.9. *Any presymplectic manifold* $(N^{2n+k},\hat{\omega})$, rank $\hat{\omega} = 2n$, *can be embedded as a coisotropic submanifold of a certain (non unique) symplectic manifold of dimension* $2(n+k)$.

Proof. Let F be the annihilator foliation of $\hat{\omega}$, and let us cover N

by an atlas of local adapted coordinates (x^u, y^a) which satisfy the transition relations (3.2.4). The notation is the same as the one used in (3.2.4). Let H be a horizontal distribution in TN, i.e., $TN = H \oplus F$. Then, there are some well defined local functions t^a_u such that $H = \text{span}\,\{B_u\}$ with

$$B_u = \frac{\partial}{\partial x^u} - t^a_u \frac{\partial}{\partial y^a} \quad . \tag{3.2.13}$$

$(\partial/\partial y^a, B_u)$ are local *adapted bases* of TN, and they have the dual cobases

$$\beta^a = dy^a + t^a_u dx^u \; , \; dx^u \quad . \tag{3.2.14}$$

It is easy to establish that the transition relations of β^a are

$$\tilde{\beta}^a = \frac{\partial \tilde{y}^a}{\partial y^b} \beta^b \quad . \tag{3.2.15}$$

Now, let $V^{2(n+k)}$ denote the total space of the dual vector bundle F^*. In V we have the local coordinates (x^u, y^a, ξ_a) where $\varphi \in F^*$ is expressed as $\varphi = \xi_a \beta^a$. We also have the natural projection $\pi : F^* \to N$. Then, the canonical symplectic form (1.1.5) of a cotangent bundle suggests us to define

$$\omega = \pi^* \hat{\omega} + d(\xi_a \beta^a) \; , \tag{3.2.16}$$

and it follows from (3.2.15) that ω is a well defined closed 2-form on V.

Furthermore, we clearly have $\operatorname{rank} \omega|_{\xi_a = 0} = 2(n+k)$, hence the same is true in some neighbourhood V^0 of the 0-section of F^* in V. It follows that (V^0, ω) is a symplectic manifold, where N is embeeded as the zero section of F^*, and then $\hat{\omega}$ is induced by ω. This embedding is easily seen to be coisotropic. Q.e.d.

A class of presymplectic manifolds which is of a particular

interest is given by

Definition 3.2.10. *A **contact form** is a 1-form* θ *defined on a differentiable manifold* P^{2n+1}*, and such that*

$$(3.2.17) \qquad (\forall p \in P)\left(\theta_p \wedge (d\theta)_p^n \neq 0\right) .$$

In this case, the pair (P,θ) *is called a **contact manifold**. (Of course, in (3.2.17) the exponent* n *indicates an exterior power.)*

Since (3.2.17) implies $\operatorname{rank} d\theta = 2n$ everywhere, $d\theta$ is a presymplectic form on P. Contact manifolds play as well an important role in physics, where they are used, for instance, to represent dynamical systems with time dependent Hamiltonian [1]. The geometry of contact manifolds has been much investigated, and the interested reader will find in [13], for instance, an exposition and a bibliography for this subject.

Let (P^{2n+1},θ) be a contact manifold. Then the annihilator of $d\theta$ is 1-dimensional, and we may see it as the space generated by the vector E such that

$$(3.2.18) \qquad i(E)d\theta = 0 \quad , \quad i(E)\theta = 1 .$$

Since (3.2.18) has a global character, we have in fact a global vector field E, and this is called the *characteristic vector field* of the contact manifold. If the orbits of E form a nice foliation, we have a symplectic reduction P/E. Moreover, by a theorem of Boothby and Wang [13], it is known that if P is compact, and if the orbits of E form a *regular foliation* (i.e., the leaves are the fibers of a submersion [69] then P is a principal circle bundle over the symplectic manifold P/E, and θ is proportional to a connection form on E. A contact manifold (P,θ) whose foliation E is regular is called a *regular contact manifold.*

On the other hand, on any contact manifold, we have a field C of hyperplanes defined by

$$(3.2.19) \qquad \theta = 0 \quad ,$$

and we call it the *contact distribution* of P. Since $\theta(E) = 1$, the contact distribution is supplementary to the annihilator, and we may see it as a horizontal distribution on P. Accordingly, we have

Proposition 3.2.11. *If* (P,θ) *is a contact manifold, then* $(C,d\theta|_C)$ *is a symplectic vector bundle on* P.

Finally, let us quote some basic examples of contact manifolds. First comes $\mathbb{R}^{2n+1}$ with the coordinates denoted by (x^i,y^i,z) $(i = 1,\dots,n)$, where

$$(3.2.20) \qquad \theta_0 = dz - \sum_{i=1}^{n} y^i dx^i$$

is easily seen to define a contact form. More generally, the form defined by the local expressions (3.2.20) yields a contact structure on the jet manifold $J^1(M^n,\mathbb{R})$, with natural canonical coordinates, for every differentiable manifold M^n. Then, let $\iota : P^{2n+1} \to \mathbb{R}^{2n+2}$ be a hypersurface of $\mathbb{R}^{2n+2}$ such that non of its tangent planes passes through the origin; if the coordinates of $\mathbb{R}^{2n+2}$ are (x^k) $(k = 1,\dots,2n+2)$, and if we consider the 1-form

$$(3.2.21) \qquad \theta = \sum_{\alpha=1}^{n+1}\left(x^{2\alpha-1}dx^{2\alpha} - x^{2\alpha}dx^{2\alpha-1}\right) \quad .$$

then we can see by a straightforward computation, and in view of the geometric condition imposed that $\iota^*\theta_1$ is a contact form on P. Particularly, the odd-dimensional unit spheres S^{2n+1} are endowed in this way with a contact structure.

For a last example, let M^{n+1} be a riemannian manifold with metric g, and let

(3.2.22) $$\iota : S^*M = \{(x,\xi) \in T^*M / g(\xi,\xi) = 1\} \subsetneq T^*M$$

be the *unit spheres cotangent bundle* of M. Then, we can prove

Proposition 3.2.12. *The 1-form*

(3.2.23) $$\eta = \iota^*\lambda \qquad ,$$

where λ *is the Liouville form of* T^*M, *is a contact form on* S^*M.

Proof. Remember that the Liouville form is defined by (1.1.4), and here we shall use the same local coordinates of T^*M as in (1.1.4) with indices running from 1 to $n+1$. Since $\dim S^*M = \dim (T^*M)-1$, it is a co-isotropic submanifold, and $(S^*M, \iota^*d\lambda)$ is a presymplectic manifold where $\mathrm{rank}\,(\iota^*d\lambda) = 2n$. Now $g(\xi,\xi) = 1$ means $g^{hk}\xi_h\xi_k = 1$, and by derivating we get

(3.2.24) $$g^{hk}\xi_h\nabla_\ell\xi_k = 0 \quad ,$$

where ∇ is the Levi-Civita connection of g [45]. From (3.2.24) we obtain by a straightforward computation $i(A)(\iota^*d\lambda) = 0$, where

(3.2.25) $$A = g^{kh}\xi_k\left(\frac{\partial}{\partial x^h} + \Gamma^p_{hq}\xi_p\frac{\partial}{\partial \xi_q}\right) \quad ,$$

which is a tangent field of S^*M. In (3.2.25), Γ are the Christoffel symbols of the metric g. (The geometric meaning of A will be explained in Section 4.5). The vector field A spans the annihilator of $\iota^*d\lambda$. But from (3.2.25) it also follows $(\iota^*\lambda)(A) = 1$, hence A does not belong to the distribution $\iota^*\lambda = 0$ of S^*M. Now, if we take a basis of $T(S^*M)$ that consists of A and of vectors which satisfy $\iota^*\lambda = 0$, we may check immediately that $\iota^*\left(\lambda \wedge (d\lambda)^n\right) \neq 0$ everywhere. Q.e.d.

Remark 3.2.13. *It is also possible to construct a contact structure on the tangent unit spheres bundle of* M [13].

Many other geometric structures with symplectic bundles are also of interest. So are the almost contact metric structures [13], the metric f-structures [89], where the symplectic bundle is associated with a so-called fundamental 2-form, the general Jacobi and Poisson structures [55], where the symplectic structure is on the cotangent bundle of the manifold, and it may have singularities, etc. We shall not discuss these structures here, and we send the reader to the papers quoted above for details.

3.3. LAGRANGIAN SUBBUNDLES. The aim of this Section is to discuss some natural occurrences of Lagrangian subbundles in the symplectic bundles discussed in Section 3.2.

We explained in example 1.1.5a) how to attach a Lagrangian subbundle to a Lagrangian submanifold N of a symplectic manifold (V^{2n},ω), and the same is true for almost symplectic manifolds. Accordingly, we shall discuss the basic examples of Lagrangian submanifolds.

Consider a cotangent bundle T^*N, with the canonical symplectic form of (1.1.5), and a function $f : N \to \mathbb{R}$. Then $df : N \to T^*N$, defines an embedding such that $(df)^*\Omega = 0$. Hence, N is embedded by df as a Lagrangian submanifold of T^*N.

On the other hand, let X^h be an arbitrary submanifold of N. Then, $[(TN|_X)/TX]^*$ is a subbundle of rank $n-h$ of $T^*N|_X$. Its total space is an n-dimensional submanifold Y of T^*N, and if we use for X local equations $x^\sigma = 0$ $(\sigma = 1,\dots,n-h)$, we see immediately that the Liouville form λ of (1.1.4) vanishes on Y. Accordingly, Y is a Lagrangian submanifold of T^*N.

Let $\Delta(\mathbb{R}^n) \approx T^*S^{n-1}$ be the symplectic manifold of the oriented straight lines of $\mathbb{R}^n$. Let Σ be an h-dimensional submanifold of $\mathbb{R}^n$, and $\Delta_n(\Sigma)$ be the set of the oriented normals of Σ. Since a normal line of Σ is identified by its contact point of radius vector r, and by its unit vector x_0, and since the normal space of Σ at $x \in \Sigma$ is (n-h)-dimensional, we see that $\Delta_n(\Sigma)$ is an $h + (n-h-1) = (n-1)$-dimensional submanifold of $\Delta(\mathbb{R}^n)$. Moreover, the normality condition means $x_0 dr = 0$ (scalar product), whence $dx_0 \wedge dr = 0$, and by looking at the symplectic form Ω of $\Delta(\mathbb{R}^n)$ as described at the beginning of Section 3.2, we see that $\Delta_n(\Sigma)$ is a Lagrangian submanifold of $\Delta(\mathbb{R}^n)$.

Another particular case of a cotangent bundle is $\mathbb{R}^{2n} = T^*\mathbb{R}^n$, where the canonical coordinates (x^i, ξ_i) $(i=1,\ldots,n)$ of (1.1.4) are defined globally. In this case, the Lagrangian submanifolds are characterized by

Proposition 3.3.1. *Let* L *be an n-dimensional submanifold of* $\mathbb{R}^{2n}$. *Then* L *is a Lagrangian submanifold iff every point* $x_0 \in L$ *has a neighbourhood in* L *which is defined in* $\mathbb{R}^{2n}$ *by explicit equations of the form*

$$x^{\bar{I}} = x^{\bar{I}}(x^I, \xi_{\bar{I}}), \quad \xi_I = \xi_I(x^I, \xi_{\bar{I}}) \quad , \tag{3.3.1}$$

where $I \subset \{1,2,\ldots,n\}$, *and* $\bar{I} = \{1,2,\ldots,n\} \setminus I$, *and*

$$\begin{gathered} {}^t\left(\frac{\partial x^{\bar{I}}}{\partial \xi_{\bar{I}}}\right) = \left(\frac{\partial x^{\bar{I}}}{\partial \xi_{\bar{I}}}\right), \quad {}^t\left(\frac{\partial \xi_I}{\partial x^I}\right) = \left(\frac{\partial \xi_I}{\partial x^I}\right), \\ \left(\frac{\partial \xi_I}{\partial \xi_{\bar{I}}}\right) = -\left(\frac{\partial x^{\bar{I}}}{\partial x^I}\right) \quad . \end{gathered} \tag{3.3.2}$$

Proof. Here, (3.3.2) are matrix relations where the indices of the entries have to run in the sets I, $\bar{I}$ as indicated. It is immediate that (3.3.1) and (3.3.2) imply $dx^i \wedge d\xi_i = 0$, hence the submanifold is Lagrangian in $\mathbb{R}^{2n}$.

Conversely, let $x_0 \in L$, a Lagrangian submanifold, and let us consider the linear subspace $A = T_{x_0}L \cap \mathrm{span}\,\{\partial/\partial x^i\}_0$ and assume $s = \dim A$. Then A has a basis $v_h = \alpha_h^i(\partial/\partial x^i)$, where $\mathrm{rank}\,(\alpha_h^i) = s$, hence it also has a basis of the form

$$(3.3.3)\qquad w_h = \frac{\partial}{\partial x^h} + \beta_h^u \frac{\partial}{\partial x^u} \qquad ,$$

where $h \in I = \{i_1,\dots,i_s\} \subset \{1,\dots,n\}$, $u \in \bar{I}$.

Let us define

$$(3.3.4)\qquad B = \mathrm{span}\left\{\frac{\partial}{\partial x^h} + \beta_h^u \frac{\partial}{\partial x^u}\ ,\ \frac{\partial}{\partial \xi_u} - \beta_h^u \frac{\partial}{\partial \xi_h}\right\}_0 \qquad .$$

Then B is a Lagrangian subspace, and

$$(3.3.5)\qquad B \cap \mathrm{span}\left\{\frac{\partial}{\partial x^i}\right\}_0 = T_{x_0}L \cap \mathrm{span}\left\{\frac{\partial}{\partial x^i}\right\}_0$$

Hence, by Proposition 2.2.18, and Remark 2.2.19, there is a linear symplectic transformation σ of $\mathbb{R}^{2n}$ which sends B onto $T_{x_0}L$, and it has a matrix of the form (2.2.13), with respect to the basis (3.3.4) and its dual. It follows that $T_{x_0}L$ has a basis of the form

$$(3.3.6)\qquad \left\{\frac{\partial}{\partial x^h} + \beta_h^u \frac{\partial}{\partial x^u}\ ,\ \frac{\partial}{\partial \xi_u} - \beta_h^u \frac{\partial}{\partial \xi_h} + \gamma_u^i \frac{\partial}{\partial x^i}\right\}_0 \qquad ,$$

which is the image of the basis (3.3.4) by σ. In (3.3.6), γ_u^i are some uninteresting coefficients, and the indices take the same values as in (3.3.3).

Furthermore, it follows from (3.3.6) that $dx^h|_0, d\xi_u|_0$ $(h \in I, u \in \bar{I})$ are linearly independent covectors on L, and, hence, (x^h, ξ_u) may be taken as local coordinates on some neighbourhood of x_0 in L. This yields (3.3.1), and then (3.3.2) follow from the Lagrangian character of L. Q.e.d.

Corollary 3.3.2. *A Lagrangian submanifold* L *of* $\mathbb{R}^{2n}$ *has a canonical atlas* $\{U_I, (x^h, \xi_u)_{(h\in I, u\in \bar{I})}\}$ *where* I *runs over subsets of* $\{1,\ldots,n\}$.

Let (V_a^{2n}, ω_a) (a=1,2) be two symplectic manifolds, and $f: V_1 \to V_2$ a mapping such that $\omega_1 = f^*\omega_2$. Such a mapping is called a *canonical transformation*, and it is necessarily a local diffeomorphism (if diffeomorphic, f is a symplectomorphism). Then $(V_1 \times V_2, p_1^*\omega_1 - p_2^*\omega_2)$, where p_1, p_2 are the projections of the direct product, is a symplectic manifold, and

$$\iota : \text{graph } f = \{(x, f(x))/x \in V_1\} \subsetneq V_1 \times V_2 \tag{3.3.7}$$

is a Lagrangian submanifold embedded in $V_1 \times V_2$.

This example suggests the following definition which proved to be very useful in Hamiltonian physics [1]

Definition 3.3.3. *Let* $(V_a^{2n_a}, \omega_a)$ (a = 1,2) *be symplectic manifolds. A Lagrangian submanifold of* $(V_1 \times V_2, p_1^*\omega_1 - p_2^*\omega_2)$ *is called a* ***canonical relation*** *between* V_1 *and* V_2.

Lagrangian submanifolds can also be obtained by the reduction procedure of [87], [1] (see Proposition 2.2.11).

Definition 3.3.4. *Two submanifolds* N_1, N_2 *of a manifold* N *are said to have a* ***clean intersection*** *if* $N_1 \cap N_2$ *is again a submanifold, and* $T(N_1 \cap N_2) = TN_1 \cap TN_2$.

Remark 3.3.5. *If* N_1 *and* N_2 *intersect transversally (i.e.,* $TM = TN_1 + TN_2$*) then, by known transversality results* [40] N_1, N_2 *have a clean intersection.*

Now, we can prove [87]

Theorem 3.3.6. *Let* (N^k, ω_N) *be a coisotropic submanifold with a nice annihilator foliation of a symplectic manifold* (V^{2n}, ω), *and let* L^n *be a Lagrangian submanifold of* V. *If* L,N *have a clean intersection, then* $L \cap N$ *projects to an immersed Lagrangian submanifold of the reduced symplectic space of* N.

Proof. N is of constant rank 2(k-n), hence, by Corollary 3.2.5, it has a 2(k-n)-dimensional symplectic reduction N/F, where F is the annihilator of ω_N, and, since ω_N is induced by ω, $F = {}^{s}TN$. Let $\pi : N \to N/F$ denote the natural projection and π_* be its differential. Then, for $x \in L \cap N$, we have the symplectic vector space $T_x V$, the coisotropic subspace $T_x N$, the Lagrangian subspace $T_x L$, and the symplectic reduction $T_{\pi(x)}(N/F)$. $\pi_*(x)$ is the projection on this symplectic reduction, and, by Proposition 2.2.11, $\pi_*(TL \cap TN) = \pi_*\{T(L \cap N)\}$ is a Lagrangian subbundle of $T(N/F)$. Implicitly, rank $\pi_*(x)$ = dim im $\pi_*(x)$ = k-n = constant. The assertion of the Theorem is now clear. Generally, we have an immersed submanifold only since $L \cap N$ may intersect several times a leaf of F. Q.e.d.

Remark 3.3.7. *In Theorem 3.3.6, if* L *is transversal to* N , *i.e.,* $TL + TN = TV$, *we get easily* $\dim (L \cap N) = k-n$, *and* π_* *is a local diffeomorphism.*

Corollary 3.3.8. [87]. *Let* (V_a, ω_a) $(a = 1,2,3)$ *be symplectic manifolds, and* $C_1 \subset V_1 \times V_2$, $C_2 \subset V_2 \times V_3$ *be canonical relations. Define the composed relation*

$$C_3 = C_2 \circ C_1 = \{(x_1, x_3) \in V_1 \times V_3 / \exists x_2 \in V_2, (x_1, x_2) \in C_1, (x_2, x_3) \in C_2\} \quad . \tag{3.3.8}$$

Denote by Δ_{V_2} *the diagonal of* $V_2 \times V_2$, *and set* $\Delta = V_1 \times \Delta_{V_2} \times V_3$. *Then, if* $C_1 \times C_2$ *and* Δ *have a clean intersection,* C_3 *is also a canonical relation.*

Proof. We have the obvious identification

$$C_3 \approx pr_{V_1\times V_3}\{(C_1\times C_2)\cap\Delta\subset V_1\times V_2\times V_2\times V_3\} \qquad (3.3.9) \quad ;$$

$V = V_1 \times V_2 \times V_2 \times V_3$ can be seen as a symplectic manifold with the 2-form $\omega = p_1^*\omega_1 - p_2^*\omega_2 + p_3^*\omega_2 - p_4^*\omega_3$, where p_σ ($\sigma = 1,2,3,4$) are the four projections of the product considered. $C_1 \times C_2$ is a Lagrangian submanifold of V, and Δ is a coisotropic submanifold with a nice annihilator foliation whose leaves are $\{v_1\} \times \Delta_{V_2} \times \{v_3\}$ ($v_1 \in V_1, v_3 \in V_3$). Now, the result follows straightforwardly from Theorem 3.3.6. Q.e.d.

Remark 3.3.9. *If* V_1 *is a point and* $\omega_1 = 0$, *a canonical relation of* V_1 *and* V_2 *is* $C_1 = L =$ *a Lagrangian submanifold of* V_2, *and* $C_2 \circ C_1$ *is the image* $C_2(L) \subset V_3$. *In this case, Corollary 3.3.8 yields conditions for* $C_2(L)$ *to be a Lagrangian submanifold. Similarly, if* V_3 *is a point,* C_2 *is a Lagrangian submanifold* $L \subset V_2$, *and* $C_2 \circ C_1 = C_1^{-1}(L)$.

As an application of this remark, we have the following interesting results [37]

Proposition 3.3.10. *Let* $f : M_1 \to M_2$ *be a differentiable mapping, and let* L *be a Lagrangian submanifold of* T^*M_1. *If* rank f = const., *and if* L *and* $f^*(T^*M_2)$ *are transversal submanifolds of* T^*M_1, *then the direct image* $(f^*)^{-1}(L)$ *is a Lagrangian submanifold of* T^*M_2.

Proof. In this proposition, the cotangent bundles have the canonical symplectic structures, f^* is induced by f on covectors, and the transversality condition is meaningful because rank f = const. We will apply Corollary 3.3.8. with V_1 = point, $V_2 = T^*M_1$, $V_3 = T^*M_2$, $C_1 = L$, and

$$C_2 = \{[(x,\xi),(y,\eta)]/y = f(x),\ \xi = f^*(\eta)\} \qquad (3.3.10) \quad .$$

Then, $C_2 \circ C_1 = (f^*)^{-1}(L)$. The manifold V encountered in the proof of

Corollary 3.3.8 is $V = T^*M_1 \times T^*M_1 \times T^*M_2$, and it has local canonical cotangent coordinates (Section 1.1) $(x^i, \xi_i, y^j, \eta_j, z^\alpha, \zeta_\alpha)$ $(i,j = 1,\ldots,\dim M_1;$ $\alpha = 1,\ldots,\dim M_2)$. $C_1 \times C_2 = L \times C_2$ has the local equations

$$(3.3.11) \qquad x^i = x^i(t^h),\ \xi_i = \xi_i(t^h),\ z^\alpha = f^\alpha(y^j),\ \eta_j = \frac{\partial f^\alpha}{\partial y^j}\zeta_\alpha ,$$

where the first two equations define $L \subset T^*M_1$, and the functions f^α are the local expression of f. Furthermore, $\Delta = \Delta_{T^*M_1} \times T^*M_2$ has the local equations

$$(3.3.12) \qquad y^i = x^i,\ \eta_i = \xi_i \quad .$$

Using (3.3.11), (3.3.12) we can compute spanning vectors of the tangent spaces at $p \in (L \times C_2) \cap \Delta$, and we get

$$(3.3.13) \qquad T_p(L \times C_2) + T_p\Delta = \mathrm{span}\left\{\frac{\partial}{\partial x^i} + \frac{\partial}{\partial y^i},\ \frac{\partial}{\partial \eta_i} + \frac{\partial}{\partial \xi_i},\ \frac{\partial}{\partial z^\alpha},\ \frac{\partial}{\partial \zeta_\alpha},\right.$$
$$\left.\frac{\partial x^i}{\partial t^h}\frac{\partial}{\partial x^i} + \frac{\partial \xi_i}{\partial t^h}\frac{\partial}{\partial \xi_i},\ \frac{\partial}{\partial y^i} + \zeta_\alpha \frac{\partial^2 f^\alpha}{\partial y^i \partial y^h}\frac{\partial}{\partial \eta_h},\ \frac{\partial f^\alpha}{\partial y^k}\frac{\partial}{\partial \eta_k}\right\} .$$

On the other hand $L \subset T^*M_1$ is given by the first two equations (3.3.11), and $f^*(T^*M_2)$ is given in T^*M_1 by

$$(3.3.14) \qquad \xi_i = \frac{\partial f^\alpha}{\partial x^i}\xi_\alpha \quad .$$

This implies that, at $q \in L \cap f^*(T^*M_2)$, one has

$$(3.3.15) \qquad T_qL + T_q\left(f^*(T^*M_2)\right) = \mathrm{span}\left\{\frac{\partial x^i}{\partial t^h}\frac{\partial}{\partial x^i} + \right.$$
$$\left. + \frac{\partial \xi_i}{\partial t^h}\frac{\partial}{\partial \xi_i},\ \frac{\partial}{\partial x^i} + \xi_\alpha \frac{\partial^2 f^\alpha}{\partial x^i \partial x^j}\frac{\partial}{\partial \xi_j},\ \frac{\partial f^\alpha}{\partial x^i}\frac{\partial}{\partial \xi_\alpha}\right\} .$$

In view of (3.3.12), the relations (3.3.13) and (3.3.15) show that if L is transversal to $f^*(T^*M_2)$ then $L \times C_2$ is transversal and has a clean intersection with Δ. Hence, the conclusion follows by Corollary 3.3.8. Q.e.d.

Proposition 3.3.11. *Let* $f : M_1 \to M_2$ *be a differentiable mapping, and let* L *be a Lagrangian submanifold of* T^*M_2. *If the mappings* f *and* $\varphi = \pi_2|_L : L \to M_2$ $(\pi_2 : T^*M_2 \to M_2)$ *are transversal mappings, the inverse image* $f^*(L)$ *is a Lagrangian submanifold of* T^*M_1.

Proof. This time, we take $V_1 = T^*M_1$, $V_2 = T^*M_2$, V_3 = point, $C_2 = L$, and C_1 defined by formula (3.3.10). Then, $C_2 \circ C_1 = f^*(L)$. The manifold V of the proof of Corollary 3.3.8 is $V = T^*M_1 \times T^*M_2 \times T^*M_2$, and it has local canonical cotangent coordinates $(x^i, \xi_i, y^\alpha, \eta_\alpha, z^\alpha, \zeta_\alpha)$ $(i=1,\ldots,\dim M_1;$ $\alpha=1,\ldots,\dim M_2)$. Accordingly, the local equations of $C_1 \times C_2$ are of the form

$$(3.3.16) \qquad y^\alpha = f^\alpha(x^i),\ \xi_i = \frac{\partial f^\alpha}{\partial x^i}\eta_\alpha,\ z^\alpha = z^\alpha(t^\sigma),\ \zeta_\alpha = \zeta_\alpha(t^\sigma)\ ,$$

where the last two equations are those of $L \subset T^*M_2$. The local equations of $\Delta = T^*M_1 \times \Delta_{T^*M_2}$ are

$$(3.3.17) \qquad z^\alpha = y^\alpha,\ \zeta_\alpha = \eta_\alpha \quad .$$

Then, a straightforward computation shows that for $p \in (C_1 \times C_2) \cap \Delta$ one has

$$(3.3.18) \qquad T_p(C_1 \times C_2) + T_p\Delta = \text{span}\left\{\frac{\partial}{\partial x^i}, \frac{\partial}{\partial \xi_i}, \frac{\partial}{\partial y^\alpha} + \frac{\partial}{\partial z^\alpha}, \frac{\partial}{\partial \eta_\alpha}, \frac{\partial}{\partial \zeta_\alpha}, \frac{\partial f^\alpha}{\partial x^i}\frac{\partial}{\partial y^\alpha}, \frac{\partial z^\alpha}{\partial t^\sigma}\frac{\partial}{\partial t^\alpha}\right\} .$$

Since such p satisfies $z^\alpha = y^\alpha$, it follows that the tranversality hypothesis of the proposition implies the transversality, hence the clean intersection, of $C_1 \times C_2$ and Λ. Then, Corollary 3.3.8 implies the result. Q.e.d.

In connection with the Lagrangian submanifolds let us also mention in passing the following interesting problem. If (V^{2n}, ω) is a symplectic manifold, then what are the n-dimensional manifolds N^n that admit a Lagrangian immersion or embedding in V? More specific, if V^{2n}, N^n, and a mapping $j : N \to V$ are given, does there exist a Lagrangian immersion $i : N \to V$ homotopic to j? The following conditions are obviously necessary ones: a) the de Rham cohomology class $j^*[\omega] = 0$; b) there is an injective bundle mapping from TN onto a Lagrangian subbundle of the pull back $j^*(TV)$, or, equivalently, $j^*(TV) \approx TN \oplus T^*N$. It has been shown by Gromov [35] and Lees [48] that these conditions are also sufficient.

Now, let us refer to contact manifolds (P^{2n+1}, θ) (Definition 3.2.10), and to the symplectic bundle provided by the contact distribution C $(\theta = 0)$ (Proposition 3.2.11). Because of (3.2.17) C is never completely integrable, but it may have integral submanifolds Λ^h of dimension $h < 2n$.

Proposition 3.3.12. *If* Λ^h *is a submanifold of* P *such that* $T\Lambda \subset C$, *then* $T\Lambda$ *is an isotropic subbundle of* C.

Proof. Let X, Y be vector fields on Λ, and implicitly $\theta(X) = \theta(Y) = 0$. Then $[X,Y] \in T\Lambda$, hence we also have $\theta([X,Y]) = 0$. This implies $d\theta(X,Y) = 0$. Q.e.d.

From this we deduce that in fact $h \leqslant n$, and we give

Definition 3.3.13. *An n-dimensional integral submanifold* Λ^n *of* C *is called a* ***Legendrian submanifold*** *of* P.

Corollary 3.3.14. *If* Λ *is a Legendrian submanifold of* P *then* $T\Lambda$ *is a Lagrangian subbundle of the symplectic vector bundle* C/Λ.

This is another basic example of a Lagrangian subbundle.

Let us also give a few concrete examples of Legendrian submanifolds. If P is $\mathbb{R}^{2n+1}$ with the contact form θ_0 of (3.2.20), then the planes $x^i = \text{const.}$ $z = \text{const.}$ are, obviously, Legendrian submanifolds. If $f : M^n \to \mathbb{R}$ is a function, then the cross section of its 1-jets is a Legendrian submanifold of the contact manifold $J^1(M,\mathbb{R})$ (see again formula 3.2.20)). Now, let us consider the unit spheres cotangent bundle S^*M of (3.2.22), and let X^h be a submanifold of M. Let us define

$$Y = [(TM/X)/TX]^* \cap (S^*M/X) \quad . \tag{3.3.19}$$

Then, it follows again that the Liouville 1-form vanishes on Y (see the example at the beginning of this Section), and Y is a Legendrian submanifold of S^*M.

It is important to note that the fibers of S^*M are also Legendrian submanifolds. They define a *vertical foliation* C of S^*M (similar to V on T^*M; see Section 1.1) which is a *Legendrian foliation* i.e., its leaves are Legendrian submanifolds.

Remark 3.3.15. *It is clear that examples of Lagrangian subbundles are also provided by tangent bundles of either Lagrangian foliations of symplectic manifolds or Legendrian foliations of contact manifolds. Important results have been established for Lagrangian foliations* [86], [53], *etc., but we shall not develop this theory here.*

We shall end this Section by touching a question which is important for both this and the previous Sections, namely, the compatible complex structures of the symplectic vector bundles of Section 3.2.

By Theorem 3.1.2, every almost symplectic manifold (V,ω) has associated compatible almost complex structures J and hermitian metrics g such that (V,J,g) is an *almost hermitian manifold* [45]. Conversely, every almost hermitian manifold has an almost symplectc structure which is defined by the Kähler form of the manifold. In the symplectic case $d\omega = 0$, hence we have

Proposition 3.3.16. *Every symplectic manifold admits compatible almost Kählerian structures.*

It is well known that the most important almost Kählerian manifolds are the *Kähler manifolds* [45]. However, examples of symplectic manifolds that do not admit any Kähler metric (i.e., for any positive compatible J, the latter is nonintegrable) exist even in the compact case [78], and even in the compact simply connected case [27]. Since these examples are of a relatively recent date, let us quote the following one due to Thurston [78]. Consider the manifold $M_4 = \mathbb{R}^4/G$, where G is the group of transformations

$$(3.3.20)\qquad (x^1,x^2,x^3,x^4) \mapsto (x^1 + p, x^2 + q, x^3 + qx^4 + m, x^4 + n) \quad ,$$

where p,q,m,n are integers. M_4 is a compact symplectic manifold with the universal covering $\mathbb{R}^4$, and the symplectic form induced by

$$(3.3.21)\qquad \omega = dx^1 \wedge dx^2 + dx^3 \wedge dx^4 \quad .$$

The fundamental group $\pi_1(M)$ is isomorphic to G, and it is easy to see that its commutator is the subgroup obtained by taking $p=0$, $q=0$, $n=0$ in (3.3.20). Accordingly, the first homology group of M is

$$(3.3.22)\qquad H_1(M) \approx \pi_1(M)/_{[\pi_1(M),\pi_1(M)]} \approx \mathbb{Z} \oplus \mathbb{Z} \oplus \mathbb{Z} \ ,$$

and the first Betti number is $b_1(M) = 3$. This contradicts the well-known fact that compact Kähler manifolds have an even first Betti number [31]. Hence M has no Kähler metrics. This example is generalized in [20]. The simply connected case is more complicated [27].

Now, let N be a Lagrangian submanifold of (V,ω). Then, if a choice of J is made, the structure group of the symplectic bundle TV/N

is reduced to $O(n)$ as in Proposition 3.1.6, while the metric g induces a riemannian metric of N. Let us also restate the following fact

Proposition 3.3.17. *If* N *is a Lagrangian submanifold of* (V,ω), *it is possible to associate positive compatible complex structures of* TV/N *with every riemannian metric* γ *of* N *such that the corresponding metric* g *induces* γ *on* TN.

Proof. Take structures of the form $J_{(TN,L',\gamma)}$ as encountered in the proof of Theorem 2.3.3. Q.e.d.

Furthermore, let (P^{2n+1},θ) be a contact manifold, and C be its contact distribution. Then any tangent vector field v of P can be expressed uniquely as

$$v = v' + \theta(v)E \ , \quad v' \in C. \tag{3.3.23}$$

Let J_C be a compatible complex structure of $(C,d\theta_C)$, and define $\Phi : TP \to TP$ by

$$\Phi v = J_C v' \ . \tag{3.3.24}$$

Then, Φ satisfies the relations

$$\Phi^2 = -Id. + \theta \otimes E, \ \Phi(E) = 0, \ \theta\circ\Phi = 0, \tag{3.3.25}$$

and it is what one calls an *almost contact structure tensor* of P [13]. The tensor Φ also satisfies the compatibility condition

$$d\theta(\Phi v,\Phi w) = d\theta(v,w) \ , \tag{3.3.26}$$

which follows from the compatibility of J_C, and (3.2.18). Conversely, if Φ is given, and it satisfies (3.3.25), (3.3.26), we obtain J_C by (3.3.24).

Let g_C be the hermitian metric of C associated to J_C. Then

(3.3.27) $$g(v,w) = g_C(v',w') + \theta(v)\theta(w) \quad ,$$

yields a Riemannian metric of P such that

(3.3.28) $$\begin{aligned} g(\Phi v,\Phi w) &= g(v,w) - \theta(v)\theta(w) \quad , \\ d\theta(v,w) &= g(\Phi v,w) \quad . \end{aligned}$$

Hence J_C yields a *contact metric* on P [13].

Remark 3.3.18. *It is possible to formulate an analogon of Proposition 3.3.17 for Legendrian submanifolds.*

Similar constructions leading to *metric* f-*structures* [89] can be made on more general presymplectic manifolds but then a horizontal space supplementary to the annihilator of the presymplectic form has to be chosen.

3.4. LOCAL EQUIVALENCE THEOREMS IN SYMPLECTIC GEOMETRY. There is quite a number of remarkable local equivalence theorems in symplectic geometry, which provide normal (canonical) forms of various geometric structures: symplectic manifolds, submanifolds, foliations, etc. Since these are principially important theorems, we shall prove some of them in this Section. In this context, the word local was used with two meanings: i) a neighbourhood of a point; ii) a tubular neighbourhood of a submanifold. Hereafter, we shall distinguish between these two meanings by using the term *local* for i) and *tubular* for ii). The basic local theorems have been established long ago by Darboux, Lie, Cartan, Carathéodory, while the tubular theorems have been discovered in the 1970's by A. Weinstein [86], who generalized a method of J. Moser [67]. As a matter of fact, the local theorems can be refound as particular cases of the tubular theorems. However, we prefer to prove some of them independently, following the recent exposition of P. Libermann [53], [54].

We need the following simple preparations [53]. Let us look at two copies of $\mathbb{R}^p$ with the coordinates (x^i), (y^i) $(i = 1,\dots,p)$ respectively. A differential form φ on $\mathbb{R}^p \times \mathbb{R}^p = \mathbb{R}^{2p}$ is said to be *bihomogeneous of type* (r,s) if

$$\varphi = \frac{1}{r!s!}\varphi_{i_1\dots i_r j_1\dots j_s}(x,y)\,dx^{i_1}\wedge\dots\wedge dx^{i_r}\wedge dy^{j_1}\wedge\dots\wedge dy^{j_s} \quad, \tag{3.4.1}$$

and these forms span the whole Grassmann algebra of $\mathbb{R}^{2p}$. Accordingly, on $\mathbb{R}^{2p}$ we have a decomposition

$$d = d_x + d_y \quad, \tag{3.4.2}$$

where the terms denote the exterior differentiation with respect to x and y alone, and they are operators of type (1,0) and (0,1). By type decomposition, $d^2 = 0$ is equivalent to

$$d_x^2 = d_y^2 = d_x d_y + d_y d_x = 0 \quad. \tag{3.4.3}$$

Moreover, d_x, d_y satisfy Poincaré lemmas: if $d_x\varphi = 0$ where φ is defined on the whole space $\mathbb{R}^{2p}$, and it has the type (r,s) $(r \geq 1)$, then $\varphi = d_x\psi$ where ψ is again defined on $\mathbb{R}^{2p}$, and it has the type $(r-1,s)$, and similarly for d_y. (The proof is like for the Poincaré lemma along y = const. or x = const. See [80], Theorem 5.21, p. 215.)

Lemma 3.4.1. *Let* J_x *be the ideal generated by the 1-forms* (dx^i) *in the Grassmann algebra of* $\mathbb{R}^{2p}$*. Let* λ *be a* k*-form* $(k \geq 1)$ *on* $\mathbb{R}^{2p}$ *such that* $d\lambda \in J_x$*. Then, there are on* $\mathbb{R}^{2p}$ *a* k*-form* $\alpha \in J_x$*, and a* $(k-1)$*-form* β *such that*

$$\lambda = d\beta + \alpha \quad. \tag{3.4.4}$$

Proof. We have $\lambda = \tilde{\lambda} + \lambda_{0k}$, where $\tilde{\lambda} \in J_x$ and λ_{0k} is the $(0,k)$-component of λ, and $d\lambda \in J_x$ iff $d_y\lambda_{0k} = 0$. Hence, by the d_y-Poincaré lemma, there is a (k-1)-form μ such that $\lambda_{0k} = d_y\mu = d\mu - d_x\mu$, and, obviously, $d_x\mu \in J_x$. Hence, we may take $\beta = \mu$, $\alpha = \tilde{\lambda} - d_x\mu$. Q.e.d.

Corollary 3.4.2. *Let* ψ *be a closed* k-*form* ($k \geq 2$) *on* $\mathbb{R}^{2p}$. *If* $\psi \in J_x$, *there are* (k-2)-*forms* $\alpha^1,\ldots,\alpha^p$ *on* $\mathbb{R}^{2p}$ *such that*

$$\psi = dx^1 \wedge d\alpha^1 + \ldots + dx^p \wedge d\alpha^p \,. \tag{3.4.5}$$

Proof. Since ψ is closed, we have $\psi = d\lambda$ where $d\lambda \in J_x$. Hence, by (3.4.4), and since $\alpha \in J_x$, we may write

$$\lambda = d\beta - dx^1 \wedge \alpha^1 - \ldots - dx^p \wedge \alpha^p \tag{3.4.6}$$

for some components $\alpha^1,\ldots,\alpha^p$. Then, (3.4.5) follows by a differentiation. Q.e.d.

Furthermore, we need another classical result which is known in analytical mechanics as the *Jacobi-Carathéodory Theorem*. In order to formulate it, we give

Definition 3.4.3. *Let* (V^{2n},ω) *be a symplectic manifold, and* f_a ($a = 1,\ldots,h$) *real (local) functions such that the Poisson brackets* $\{f_a, f_b\} = 0$. *Then* $\{f_a\}$ *is said to be a* ***system of functions in involution.***

Theorem 3.4.4. (Jacobi-Carathéodory). *Let* (V,ω) *be a symplectic manifold, and* $x_0 \in V$. *If* f_a ($a = 1,\ldots,h$) *is a system of functionally independent functions in involution on a neighbourhood* U *of* x_0, *then* $h \leq n$, *and if* $h < n$, *the system can be completed by* $f_{h+1},\ldots,f_n$ *to a system of* n *functionally independent functions in involution on a certain neighbourhood* U', $x_0 \in U' \subset U$.

Proof. For Poisson brackets, and related facts see Section 1.2. The functions f_a in question are functionally independent iff df_a are linearly independent. Now, let f_a be the given system in involution, and let us look at the local foliation N_U defined on U by f_a = const. Let X_{f_a} be the symplectic gradients of f_a defined by (1.2.2). According to (1.2.2), these vector fields are linearly independent at every point of U. If Y is a tangent field of N_U, we have

$$\omega(X_{f_a}, Y) = df_a(Y) = 0 , \tag{3.4.7}$$

and it follows that X_{f_a} span the skeworthogonal distribution ${}^sTN_U$. On the other hand, we have

$$\{f_a, f_b\} = X_{f_a} f_b = df_b(X_{f_a}) = 0 , \tag{3.4.8}$$

whence $X_{f_a} \in TN_U$. It follows that N_U is a coisotropic foliation of codimension h, and necessarily $h \leqslant n$.

Furthermore, ${}^sTN_U$ is the annihilator of ω_{N_U}, and it follows like in the proof of Proposition 3.2.3 that ${}^sTN_U$ is an h-dimensional foliation of U, which is a subfoliation of N_U. If $h < n$, there will be at least one more first integral g of ${}^sTN_U$ defined on a neighbourhood U', $x_0 \in U' \subset U$, independent of the first integrals f_a. This function g satisfies

$$dg(X_{f_a}) = X_{f_a} g = \{f_a, g\} = 0 .$$

Hence $f_1, \ldots, f_h$, $f_{h+1} = g$ is again a system of functionally independent functions in involution. The same procedure can be continued until we arrive at f_n. Q.e.d.

Now, we are ready for the proof of the classical *Darboux theorem*

Theorem 3.4.5. (Darboux). *Let* ω *be a closed 2-form of* rank 2p *defined on a differential manifold* M^m. *Then, every point* $x \in M$ *has a coordinate neighbourhood* U *with local coordinates* (x^i, y^i, z^σ) (i = 1,...,p;

$\sigma = p+1,\ldots,m)$ *such that*

$$\psi/U = \sum_{i=1}^{p} dx^i \wedge dy^i \quad . \tag{3.4.9}$$

Proof. (M,ω) is a presymplectic manifold, and, by Proposition 3.2.3, there is a neighbourhood U of x with local coordinates (t^s, z^σ) $(s = 1,\ldots,2p;\ \sigma = p+1,\ldots,m)$ such that z^σ are coordinates along the leaves of the annihilator foliation of ω, and t^s are coordinates on the local 2p-dimensional symplectic reduction manifold N of U. Proposition 3.2.3 also shows that ψ is a form in the t^s alone, and, since N can be taken diffeomorphic to $\mathbb{R}^{2p}$, we may apply Corollary 3.4.2.

Namely, take $x^1 = t^1$, and complete by $x^2,\ldots,x^p$, up to a maximal system of functionally independent functions in involution as shown in Theorem 3.4.4. (Restrict N conveniently, if necessary.) Then $x^1 = \text{const.},\ldots,x^p = \text{const.}$ is a Lagrangian foliation of N, and this implies $\psi \in J_x$, where J_x is defined as in Lemma 3.4.1. Then, $x^1,\ldots,x^p$ can be used as the first p coordinates in $N \approx \mathbb{R}^{2p}$, and Corollary 3.4.2 ensures us of the existence of other p functions $y^1,\ldots,y^p$ such that (3.4.9) holds good. Moreover, (x^i, y^i) $(i = 1,\ldots,p)$ have to be functionally independent since, otherwise we would contradict $\operatorname{rank}\psi = 2p$. Q.e.d.

Corollary 3.4.6. *Any two symplectic manifolds* (V_a, ω_a) $(a = 1,2)$ *of the same dimension* $2n$ *are locally symplectomorphic.*

Proof. We must have neighbourhoods U_a of two arbitrary points $m_a \in V_a$ $(a = 1,2)$, where

$$\omega_a = \sum_{i=1}^{2n} dx_a^i \wedge dy_a^i \quad , \tag{3.4.10}$$

and $(x_1^i, y_1^i) \mapsto (x_2^i, y_2^i)$ is a symplectomorphism. Q.e.d.

Remark 3.4.7. *The local coordinates* (x^i, y^i) *of a symplectic manifold* (V,ω) *defined above are called* ***canonical coordinates.*** *Thereby,* V *is covered by a* ***canonical atlas,*** *and the symplectomorphisms obtained in the proof of Corollary 3.4.6 are* ***canonical transformations.*** *It is rather usual to denote canonical coordinates by* q^i, p_i, *hence* $\omega = dq^i \wedge dp_i$, *as in analytical mechanics.*

The Darboux' theorem is the first local equivalence theorem of symplectic geometry, and formula (3.4.10) yields the *local normal form* of a symplectic manifold. Corollary 3.4.6 is quite a remarkable fact. As a consequence of Darboux' theorem, we can also obtain the local normal form of contact manifolds.

Theorem 3.4.8. *Let* τ *be a 1-form such that* $\operatorname{rank} d\tau = 2p$ *and* $\tau \wedge (d\tau)^p \neq 0$ *everywhere on an arbitrary differentiable manifold* M^m. *Then every point* $x \in M$ *has a coordinate neighbourhood* U *with local coordinates* (z, x^i, y^i, t^σ) $(i=1,\ldots,p;\sigma=1,\ldots,m-2p-1)$

$$\tau = dz - \sum_{i=1}^{p} x^i dy^i \quad . \tag{3.4.11}$$

Proof. Since $\operatorname{rank} d\tau = 2p$, we get by Darboux' theorem

$$d\tau = \sum_{i=1}^{n} dx^i \wedge dy^i = d\left(\sum_{i=1}^{p} x^i dy^i\right)$$

for some functionally independent functions x^i, y^i defined in a neighbourhood U of x. This implies the existence of z such that (3.4.11) holds good, and z is functionally independent of x^i, y^i because of $\tau \wedge (d\tau)^p \neq 0$. t^σ is just the completion of (z, x^i, y^i) up to a local system of coordinates. Q.e.d.

Let us come back to a symplectic manifold (V^{2n},ω). Then the proof of Darboux' theorem extends immediately to

Theorem 3.4.9. (Lie). *If* f_a $(a = 1,\dots,h \leq n)$ *is a system of functionally independent functions in involution defined on a neighbourhood* U *of* $x_0 \in V$, *it is possible to add the functionally independent functions* f_σ, g_i $(\sigma = h+1,\dots,n; i = 1,\dots,n)$ *defined on a subneighbourhood* U' $(x_0 \in U' \subseteq U)$ *such that*

$$\omega/U' = \sum_{i=1}^{n} df_i \wedge dg_i \quad . \tag{3.4.12}$$

Proof. Use Theorem 3.4.4 in order to get f_σ. Then ω belongs to the ideal J_f generated by $\{df_i\}$, and the functions g_i are provided by Corollary 3.4.2. Q.e.d.

In geometric terms, Theorem 3.4.9 yields a local normal form for coisotropic submanifolds namely

Theorem 3.4.10. *Let* N *be a coisotropic submanifold of codimension* $h \leq n$ *of the symplectic manifold* (V^{2n},ω), *and* $x_0 \in N$. *Then* x_0 *has a coordinate neighbourhood* U *in* V *with the local coordinates* (x^i, y^i) $(i = 1,\dots,n)$, *where* $x^a = 0$ $(a = 1,\dots,h)$ *is a coordinate neighbourhood of* x_0 *in* N *with the local coordinates* x^σ, y^i $(\sigma = h+1,\dots,n)$ *and*

$$\omega/U = \sum_{i=1}^{n} dx^i \wedge dy^i \quad . \tag{3.4.13}$$

Proof. If $f_a = 0$ are local equations of N about x_0, it follows by considerations similar to those encountered in the proof of Theorem 3.4.4. that f_a is a system of functionally independent functions in involution. Hence, we have (3.4.12). Now, put $f_i = x^i$ and $g_i = y^i$ and we are done. Q.e.d.

Consequently, if $N^h_\alpha \subset (V^{2n}_\alpha, \omega_\alpha)$ $(\alpha = 1,2)$ are coisotropic submanifolds, and $x^0_\alpha \in N_\alpha$, then there is a symplectomorphism of a neighbourhood U_1 of x^0_1 in V_1 onto a neighbourhood U_2 of x_2 in V_2

which sends $U_1 \cap N_1$ onto $U_2 \cap N_2$. In other words, two coisotropic submanifolds of the same codimension of symplectic manifolds of the same dimension are locally symplectically equivalent.

Remark 3.4.11. *It is clear that Theorem 3.4.10, and its significance as shown above are also true for* ***coisotropic foliations****. We only have to replace there the local equations* $x^a = 0$ *by* $x^a = \text{const}$. *(We also see that, locally, a coisotropic submanifold can be embedded as a leaf of a coisotropic foliation.)*

Corollary 3.4.12. *Let* L *be a Lagrangian submanifold (or foliation) of* (V^{2n}, ω), *and* $x_0 \in L$. *Then there are local coordinates* (x^i, y^i) *about* x_0 *such that* $x^i = \text{const}$. *define a corresponding neighbourhood of* L *(or of the leaves of* L, *if* L *is a foliation), and formula (3.4.13) holds good.*

This is just Theorem 3.4.10 for $h=n$. This Corollary offers a local normal form of Lagrangian submanifolds and foliations, and it shows that, for the same n, they are all symplectically equivalent locally.

For a further generalization of the local Theorem 3.4.9, see Cartan's theorem as proven in [53] (Theorem 3.3 of [53]). It refers to submanifolds of constant rank. Here, we prefer to give the local normal form of such submanifolds latter on as a consequence of a corresponding tubular neighbourhood such as to have also an exemplification of another technique.

Now, let us go over to some tubular equivalence theorems. Again, we shall need some preparation, and we start with (e.g., [37])

Lemma 3.4.13. *Let* $\varphi_t : X \to Y$ *be a differentiable 1-parameter family of mappings, and let* ξ_t *be the tangent field of* φ_t. *Let* σ_t *be a differentiable 1-parameter family of* k*-forms on* Y. *Then, the following formula holds good*

$$(3.4.14) \qquad \frac{d}{dt}\varphi_t^*\sigma_t = \varphi_t^*\frac{d\sigma_t}{dt} + \{i(\xi_t)d\sigma_t\} + d\{i(\xi_t)\sigma_t\} \quad .$$

Proof. First, let us explain the notation used in this Lemma. A 1-parameter family of "objects" on a manifold M is a corresponding object on, say, $M \times I$ $(I = [0,1])$ (or on $M \times \mathbb{R}$). Now, remember that for a mapping $f : M \to N$ of two manifolds, a *vector field along* f is a mapping $v : M \to TN$ such that for all $m \in M$ one has $\pi\{v(m)\} = f(m)$ ($\pi : TN \to N$ is the natural projection). In our case, for every t, ξ_t is the vector field along φ_t defined by taking $\xi_t(x)$ to be the velocity vector of the path $\varphi_t(x)$ of Y at the moment t. Furthermore, if v is a field along $f : M \to N$, and σ is a k-form on N, one has by definition

$$(3.4.15) \qquad \{i(v)\sigma\}_x(w_1,\ldots,w_{k-1}) = \sigma_{f(x)}(v, f_*w_1,\ldots,f_*w_{k-1}) \qquad (x \in M),$$

and this is a (k-1)-form on M. It is this operation i which is used in (3.4.14).

Now, let us look again at an arbitrary mapping $f : M \to N$. Then, every vector field w on M provides a vector field f_*w along f, by putting

$$(3.4.16) \qquad (f_*w)(x) = f_*\{w(x)\} \qquad\qquad (x \in M) .$$

We call f_*w the *f-field defined by* w. With this notation, and if σ is a k-form on N, we claim that the following formula holds good

$$(3.4.17) \qquad L_w(f^*\sigma) = i(f_*w)d\sigma + di(f_*w)\sigma \quad ,$$

where L_w denotes the Lie derivative. The proof is like for the similar usual formula $L_w = i(w)d + di(w)$ on M namely, the two sides of the formula are derivations of the exterior algebra $f^* \Lambda N$, which agree on functions and 1-forms, and this justifies (3.4.17) in the general case.

Now, let us address Formula (3.4.14). It is clear that giving the family φ_t is equivalent with giving the function $\tilde{\varphi} : X \times I \to Y \times I$ such that

$$\tilde{\varphi}(x,t) = \{\varphi_t(x), t\} \qquad (x \in X,\ t \in I)\ . \tag{3.4.18}$$

Then, σ_t is equivalent to a k-form $\tilde{\sigma}$ on $Y\times I$, and, on the other hand it is easy to check that

$$\left(\tilde{\varphi}_* \frac{\partial}{\partial t}\right)(x,t) = \xi_t(x) + \frac{\partial}{\partial t}\{\varphi_t(x), t\} \quad . \tag{3.4.19}$$

Let us apply (3.4.17) for $f = \tilde{\varphi}$, $w = \partial/\partial t$, and the form $\tilde{\sigma}$. We get

$$L_{\frac{\partial}{\partial t}}(\tilde{\varphi}^*\tilde{\sigma}) = i(\xi_t)d\tilde{\sigma} + di(\xi_t)\tilde{\sigma} + \tag{3.4.20}$$

$$+ \tilde{\varphi}^*\left[i\left(\frac{\partial}{\partial t}\right)d\tilde{\sigma} + d\left(i\left(\frac{\partial}{\partial t}\right)\tilde{\sigma}\right)\right] \quad .$$

Since the last bracket in (3.4.20) is precisely $L_{\frac{\partial}{\partial t}}\tilde{\sigma}$, it follows that (3.4.20) modulo $dt=0$ is exactly (3.4.14). Q.e.d.

Corollary 3.4.14. *Let* $\varphi_t : X \to X$, $t \in [0,1]$, *for a differentiable family of mappings, i.e., a homotopy in* X. *Then there is an associated homotopy homomorphism* $J : \Lambda^k X \to \Lambda^{k-1} X$ *such that*

$$\varphi_1^*\sigma - \varphi_0^*\sigma = Jd\sigma + dJ\sigma \tag{3.4.21}$$

holds good for every $\sigma \in \Lambda^k X$.

Proof. If we use (3.4.14) for our φ_t, and for $\sigma_t = \sigma$, we obtain by integrating on $[0,1]$

$$\varphi_1^*\sigma - \varphi_0^*\sigma = \int_0^1 [i(\xi_t)d\sigma]dt + d\int_0^1 [i(\xi_t)\sigma]dt \quad . \tag{3.4.22}$$

Hence, the operator J defined as

(3.4.23) $$J\sigma = \int_0^1 [i(\xi_t)\sigma]dt$$

is the operator required. Q.e.d.

Furthermore, from Corollary 3.4.14, we can deduce

Proposition 3.4.15. (Relative Poincaré Lemma [87]). *Let* Y *be an embedded submanifold of* X, *and let* σ *be a closed form on* X *whose pullback to* Y *is* 0. *Then, there is a neighbourhood* U *of* Y *in* X *such that* $\sigma|_U = d\psi_U$ *for some form* ψ_U *on* U *whose pullback to* Y *is* 0.

Proof. Let us take U to be a tubular neighbourhood of Y in X [40]. Then, we have a strong retracting deformation $\varphi_t : U \to U$ such that $\varphi_t/Y = id_Y$, $\varphi_1 = id_U$, $\varphi_0 = \iota\circ r$, where r is a retraction of U to Y, and ι is the inclusion of Y in U. If we apply (3.4.21) for this φ_t, and since $\iota^* = 0$ by hypothesis, we obtain $\sigma = dJ\sigma$ on U, which is the first assertion of our proposition, with $\psi_U = J\sigma|_U$. The property $\iota^* J\sigma = 0$ is clear from (3.4.23). Q.e.d.

Remark 3.4.16. *If the stronger condition* $\sigma|_Y \overset{def}{=} \sigma|_{(TU/Y)}$ *(we shall always use a similar notation for such restrictions) holds, we obviously have also* $\psi_U|_Y = 0$, $d\psi_U|_Y = 0$.

Now, we are ready for the proof of

Theorem 3.4.17. (A. Weinstein [86]). *Let* Y *be an embedded submanifold of* X, *and let* ω_0, ω_1 *be two symplectic forms on* X *such that* $\omega_0|_Y = \omega_1|_Y$. *Then, there is a neighbourhood* U *of* Y, *and a diffeomorphism* $f : U \to U$ *such that* $f|_Y = id_Y$, *and* $f^*\omega_1 = \omega_0$.

Proof. Let us denote $\sigma = \omega_1 - \omega_0$, and consider

(3.4.24) $$\omega_t = \omega_0 + t\sigma \quad .$$

Then, σ satisfies the hypothesis of Proposition 3.4.15, and $\sigma = d\psi$ on a tubular neighbourhood U of Y in X. We also have $\sigma|_Y = 0$, hence $\psi|_Y = 0$. Since $\omega_t|_Y = \omega_0|_Y$, this form is nondegenerate for all $t \in [0,1]$, and, by continuity, this will also be true in a, perhaps smaller, neighbourhood U of Y. Hence, we have a "time dependent" vector field ξ_t defined on U by

$$(3.4.25) \qquad i(\xi_t)\omega_t = -\psi \qquad (0 \leq t \leq 1) ,$$

and satisfying $\xi_t|_Y = 0$. Time dependent vector fields can be integrated locally just like the usual vector fields since the corresponding theorem about differential equations is true (e.g., [47]). In our case, the integral curves of ξ_t are defined for all $t \in [0,1]$ at the points of Y (since $\xi_t|_Y = 0$). Hence the same is also true on a (perhaps even smaller) neighbourhood U of Y. The corresponding flow [47] is a family of diffeomorphisms $f_t : U \to U$, $0 \leq t \leq 1$, $f_0 = \text{id.}$, and ξ_t is the tangent field of this family as encountered in Lemma 3.4.13. Accordingly, Formula (3.4.14) gives

$$(3.4.26) \qquad \frac{d}{dt}(f_t^*\omega_t) = f_t^*\left(\frac{d\omega_t}{dt}\right) + i(\xi_t)d\omega_t + d\left(i(\xi_t)\omega_t\right) =$$

$$= f_t^*\sigma + 0 + f_t^*(-d\psi) = 0 .$$

(This computation is based on (3.4.24), (3.4.25), and $\sigma = d\psi$. Pay attention to the term $f_t^*(-d\psi)$ where f_t^* appears in view of the correct use of (3.4.15) for our field ξ_t.) Finally, by integrating (3.4.26) on [0,1], we get

$$(3.4.27) \qquad f_1^*\omega_1 = \omega_0 ,$$

and we see that f_1 is the diffeomorphism f needed for Weinstein's theorem. Q.e.d.

The equivalence proven in Theorem 3.4.17 can be seen under the following more general form

Corollary 3.4.18. *Let* (V_a, ω_a) $(a = 1,2)$ *be two symplectic manifolds and* $Y_a \subset V_a$ *be embedded submanifolds. Assume that there exists a diffeomorphism* $\varphi : U_1 \to U_2$ *of the open neighbourhoods* U_a *of* Y_a *in* V_a *such that* $\varphi(Y_1) = Y_2$, *and* $\varphi^*(\omega_2|_{Y_2}) = \omega_1|_{Y_1}$. *Then there also exists a diffeomorphism* $f : U'_1 \to U'_2$ *of perhaps smaller neighbourhoods* $U_a \supset U'_a \supset Y_a$ *such that* $f(Y_1) = Y_2$, *and* $f^*\omega_2 = \omega_1$ *on* U'_1.

Proof. By Theorem 3.4.17, we have a diffeomorphism ψ on some $U_1 \supset Y_1$, which is identical on Y_1, and satisfies $\psi^*(\varphi^*\omega_2) = \omega_1$. Then, we may take $f = \varphi \circ \psi$. Q.e.d.

Particularly, if Y is a point $x_0 \in V_1$, and if $V_2 = \mathbb{R}^{2n}, Y_2 = \{0\}$, $\omega_2 = \sum_{i=1}^{n} dx^i \wedge dx^{i+n}$, we can construct φ of the Corollary as follows: choose a symplectic basis $(e_i, e_{i^*} = e_{i+n})$ of $T_{x_0} V_1$; take a coordinate neighbourhood of x_0 with the local coordinates (u^i, u^{i^*}) such that $u^i(x_0) = 0$, $u^{i^*}(x_0) = 0$, and $\partial/\partial u^i|_{x_0} = e_i$, $\partial/\partial u^{i^*}|_{x_0} = e_{i^*}$ (this can be done, for instance, by means of an exponential mapping [45]; define φ by $x^i = u^i$, $x^{i^*} = u^{i^*}$. Then, the existence of f of Corollary 3.4.18 is precisely Darboux' theorem.

Furthermore, it is very important to show that the "uniqueness" Theorem 3.4.17 has a corresponding "existence" part [87].

Theorem 3.4.19. *Let* Y *be an embedded submanifold of a manifold* X, *and let* ω *be a skewsymmetric bilinear form on* $TX/_Y$, *whose restriction to* TY *is a closed form of* Y. *Then* ω *extends to a closed 2-form on a neighbourhood* U *of* Y *in* X. *Moreover, if* ω *is a symplectic form so is its extension.*

Proof. Let U be a tubular neighbourhood of Y in X, and let us take an open covering of U by neighbourhoods of the form $V_\alpha \approx A_\alpha \times \mathbb{R}^{\operatorname{codim} Y}$ where A_α are coordinate neighbourhoods on Y which are trivializing for U. Then ω/A_α extend to 2-forms ω_α on V_α, and, if g_α is a partition of unity subordinated to V_α, the form

$$\Phi_x = \sum_{x \in V_\alpha} g_\alpha(x)\omega_\alpha(x) \qquad (x \in U) \tag{3.4.28}$$

extends ω to U, and $d\Phi/Y = 0$ (use $\Sigma g_\alpha = 1$, and the hypothesis on ω). Now, if we use Proposition 3.4.15 and Remark 3.4.16 for $\sigma = d\Phi$, we obtain on U

$$d\Phi = d\psi \, , \tag{3.4.29}$$

where $\psi/Y = 0$. Then $\Omega = \Phi - \psi$ extends ω to U, and it is closed. The last assertion of the theorem is obvious by reason of continuity. Q.e.d.

Now, we shall proceed with a tubular equivalence theorem for submanifolds of constant rank

Theorem 3.4.20. (Marle [56]). *Let* N_a^k *be embedded submanifolds of constant rank of the symplectic manifolds* (V_a^{2n}, ω^a) $(a = 1,2)$ *and let* $\hat{\omega}^a$ *be the induced presymplectic forms of* N_a. *Assume the existence of a diffeomorphism* $f : N_1 \to N_2$ *such that* $f^*\hat{\omega}^2 = \hat{\omega}^1$. *Then* f *can be extended to a symplectomorphism* $\varphi : (U_1, \omega^1) \to (U_2, \omega^2)$ *of neighbourhoods* U_a *of* N_a *in* V_a *iff* f *can be lifted to a vector bundle symplectomorphism* $\hat{f} : {}^sTN_1/{}^iTN_1 \to {}^sTN_2/{}^iTN_2$.

Proof. A diffeomorphism f as in the theorem will be called a *presymplectomorphism*. The existence of f implies that $\operatorname{rank} \hat{\omega}^1 = \operatorname{rank} \hat{\omega}^2 = 2\ell$. For the notation in $\hat{f}$ see Section 3.1. Notice also that ${}^iTN_a$ are the annihilator bundles of $\hat{\omega}^a$. Furthermore, using the vector bundle version of Theorem 2.2.4 (justified in Section 3.1), we may write

(3.4.30) $$TV_a/N_a = {}^{\sigma}TN_a \oplus_s BN_a \oplus_s CN_a ,$$

where the terms are skeworthogonal symplectic subbundles of TV_a/N_a and

(3.4.31) $$BN_a \approx TN_a/{}^{i}TN_a , \quad CN_a \approx {}^{s}TN_a/{}^{i}TN_a , \quad {}^{\sigma}TN_a = {}^{i}TN_a \oplus QN_a ,$$

QN_a being an isotropic supplementary subbundle of $TN_a + {}^{s}TN_a$ in TV_a/N_a.

Now, it is clear that the existence of $\hat{f}$ is a necessary condition for the existence of φ. (Simply, $\hat{f}$ is induced by φ.)

In order to prove the sufficiency of this condition, we notice first that, since f is a presymplectomorphism, f lifts to the bundle isomorphism $f_*/{}^{i}TN_1 : {}^{i}TN_1 \to {}^{i}TN_2$. Moreover, since $QN_a \approx ({}^{i}TN_a)^*$ (see the explanations related to Theorem 2.2.4), we can also define a lift $f' : QN_1 \to QN_2$. Namely, if $q_1 \in QN_1$, we define $f'(q_1)$ by the equation

(3.4.32) $$i(f'(q_1))\omega^2 = (f^{-1})^*(i(q_1)\omega^1).$$

Furthermore, it is easy to check that

(3.4.33) $$\tilde{f} = f_*|_{{}^{i}TN_1} \oplus f' : {}^{\sigma}TN_1 \to {}^{\sigma}TN_2$$

is a vector bundle symplectomorphism. Indeed, we have

$$\omega^2(f'(q_1), f_*(a_1)) = (i(f'(q_1))\omega^2)(f_*(a_1)) =$$

$$= [(f^{-1})^*(i(q_1)\omega^1)](f_*(a_1)) = (i(q_1)\omega^1)(a_1) = \omega^1(q_1, a_1)$$

for every $q_1 \in QN_1$, $a_1 \in {}^{i}TN_1$, which is enough for our purpose since the terms of ${}^{\sigma}TN_a$ are isotropic subbundles.

Now, it is also clear that $f_*|_{BN_1} : BN_1 \to BN_2$ is a symplectomorphism, and $\hat{f}$ may be seen as a symplectomorphism $\hat{f} : CN_1 \to CN_2$. Hence the existence of f and $\hat{f}$ implies the existence of a lift $\bar{f}$ of f

which is a symplectomorphism

(3.4.34) $\bar{f} = \tilde{f} \oplus f_* \oplus \hat{f} : TV_1|_{N_1} \to TV_2|_{N_2}$.

Finally, let U_a be tubular neighbourhoods of N_a in V_a which are isomorphic to the "normal" bundles $CN_a \oplus QN_a$. Then, the existence of $\bar{f}$ of (3.4.34) implies the existence of a diffeomorphism $\psi : U_1 \to U_2$ such that $\psi(N_1) = N_2$, and $\psi_*|_{N_1} = \bar{f}$, hence $\psi^*(\omega^2/_{Y_2}) = \omega^1/_{Y_1}$. Now, the existence of the required symplectomorphism φ follows from Corollary 3.4.18. Q.e.d.

Particularly, for coisotropic, and Lagrangian submanifolds, ${}^sTN_a = {}^iTN_a$, and $\hat{f}$ is trivial and it always exists. Hence we get (see also [87])

Corollary 3.4.21. *Let* N_a *be* k*-dimensional coisotropic submanifolds of the symplectic manifolds* (V_a, ω^a) (a = 1,2). *Then any presymplectomorphism* $f : N_1 \to N_2$ *can be extended to a symplectomorphism* $\varphi : U_1 \to U_2$ *of some neighbourhoods* U_a *of* N_a *in* V_a.

Corollary 3.4.22. **(Weinstein [86], Kostant [37]).** *Let* L_a *be two Lagrangian submanifolds of the symplectic manifolds* (V_a, ω^a) (a = 1,2). *Then any diffeomorphism* $f : L_1 \to L_2$ *extends to a symplectomorphism* $\varphi : U_1 \to U_2$ *of some neighbourhoods* U_a *of* N_a *in* V_a. *Particularly, if* L *is a Lagrangian submanifold of* (V, ω), *the identification of* L *with the zero section of* T*L *extends to a symplectomorphism of some neighbourhood of* L *in* V *with a neighbourhood of the zero section in* T*L, *where the latter is endowed with its canonical symplectic structure.*

Remark 3.4.23. *Hence, the local-tubular normal models of Lagrangian submanifolds are the zero sections of the cotangent bundles.*

Now, we can also get the local normal model of a submanifold of constant rank [56].

Theorem 3.4.24. *Let* N^k *be an embedded submanifold of constant rank* 2ℓ *of the symplectic manifold* (V^{2n},ω). *Then, every point* $x_0 \in N$ *has a coordinate neighbourhood* U *in* V *with the local coordinates* $(y^a, x^u, z_a, w^\sigma)$ $[a = 1,\ldots,k-2\ell;\ u = 1,\ldots,2\ell;\ \sigma = 1,\ldots,2(n+\ell-k)]$ *such that* $N \cap U$ *has the local equations* $z_a = 0$, $w^\sigma = 0$, *and*

$$(3.4.35)\qquad \omega|_U = \sum_{a=1}^{k-2\ell} dy^a \wedge dz_a + \sum_{i=1}^{\ell} dx^i \wedge dx^{i+\ell} + \sum_{j=1}^{n+\ell-k} dw^j \wedge dw^{j+n+\ell-k} \quad .$$

Proof. Let U be a cubical coordinate neighbourhood centered at x_0, and such that $N_1 = N \cap U$ be a coordinate slice (i.e., N_1 is defined by the annulation of a part of the coordinates). Then, along N_1 we may assume that we have the local coordinates (y^a, x^u) obtained during the proof of Proposition 3.2.3, adapted to the annihilator foliation of the presymplectic structure $\hat{\omega}$ induced by ω on N_1. According to Proposition 3.2.3, the induced form $\hat{\omega}$ contains only the coordinates x^u, and, by Darboux' theorem, we may choose (x^u) such that

$$(3.4.36)\qquad \hat{\omega} = \sum_{i=1}^{\ell} dx^i \wedge dx^{i+\ell} \quad .$$

Furthermore, let us consider the manifold $V' = N_1 \times \mathbb{R}^{2n-k}$, where the coordinates of $\mathbb{R}^{2n-k}$ are denoted by (z_a, w^σ), and let us put on V' the symplectic structure

$$(3.4.37)\qquad \omega' = \sum_{i=1}^{\ell} dx^i \wedge dx^{i+\ell} + \sum_{a=1}^{k-2\ell} dy^a \wedge dz_a + \sum_{j=1}^{n+\ell-k} dw^j \wedge dw^{j+n+\ell-k} \quad .$$

Then, let N_2 be the submanifold of V' given by $z_a = 0$, $w^\sigma = 0$, which has dimension k, and rank 2ℓ. The obvious indentification $i = N_1 \to N_2$ is a presymplectomorphism, and it can be lifted to $\hat{i} : {}^{s}TN_1/{}^{i}TN_1 \to {}^{s}TN_2/{}^{i}TN_2$ since these are two trivial symplectic vector bundles of the same rank. Hence, by Theorem 3.4.20, i extends to a symplectomorphism of neighbourhoods of N_1, N_2, and we may use $(y^a, x^u, z_a, w^\sigma)$ as local coordinates on some neighbourhood of N_1 while (3.4.35) will hold good. Q.e.d.

The meaning of this theorem is as follows: let N_a be embedded submanifolds of constant rank of (V_a, ω^a) $(a = 1,2)$; if the induced presymplectic structures are equivalent, then these submanifolds are locally symplectically equivalent. In other words, and unlike the riemannian case, there is no local *extrinsic geometry* for submanifolds of a symplectic manifold. In fact, this result holds for arbitrary submanifolds, i.e. without the constant rank hypothesis. This is a result of Givental [5], and a proof is to be found in Bennequin [9].

Now, let us end by some remarks about foliations. With the notation of Theorem 3.4.24, we may notice that N_1 (i.e., a local submanifold of constant rank) embedds into the foliation $z_a = \text{const.}$, $w^\sigma = \text{const.}$, where the Poisson brackets of the local functions z_a, w^σ are constant (not all zero). The globalization of this situation is contained in [53, 54].

Definition 3.4.25. *A foliation* F *of a symplectic manifold* (V,ω) *is called complete if at every point of* V *there is a full system of functionally independent first integrals of* F, $f_1,\ldots,f_h$, *such that all the Poisson brackets* $\{f_i, f_j\}$ *are functionally dependent on* f_k $(i,j,k = 1,\ldots,h)$ *(one says that* $f_1,\ldots,f_h$ *is a* ***Poisson closed system****). If, moreover, the leaves of* F *are of constant rank, the foliation is called* ***symplectically regular****.*

It can be proven that (3.4.35) is the local normal form of a symplectically regular foliation, and this is Cartan's theorem mentioned earlier in this section [53].

Another important fact is that tubular theorems can also be

given for foliations. Particularly, one has

Theorem 3.4.26. **(A. Weinstein** [86]). *Let* L *be a Lagrangian foliation of the symplectic manifold* (V,ω), *and let* N *be an embedded Lagrangian submanifold of* V *such that* $TL/N \oplus TN = TV/N$. *Then, there is a neighbourhood* U_1 *of* N *in* V, *and a symplectomorphism* $\varphi : U_1 \to U_2$, *where* U_2 *is a neighbourhood of the zero section of* T^*N *endowed with the canonical symplectic structure of a cotangent bundle, such that* φ/N *identifies* N *to the zero section of* T^*N, *and* φ *sends the leaves of* L/U_1 *to the fibers of* T^*N.

The proof is by the same technique as for Theorem 3.4.17, with some more care in the construction of the symplectomorphism φ. We send the reader to the original paper of Weinstein [86] for details.

At the end of this section, we must emphasize that there are no global theorems similar to the local and tubular ones discussed above. It has been proven recently by Gromov [36] that $\mathbb{R}^{2n}$ has also symplectic forms which are not globally equivalent to the canonical form of Darboux' theorem.

CHAPTER 4

TRANSVERSALITY OBSTRUCTIONS OF LAGRANGIAN SUBBUNDLES (MASLOV CLASSES)

We already know that a Lagrangian subbundle of a symplectic vector bundle is equivalent to a reduction of the structure group from $Sp(n,\mathbb{R})$ to $O(n)$, via $U(n)$, and, accordingly, we shall have orthogonal connections. If two Lagrangian subbundles are given, we have two classes of orthogonal connections, whose comparison leads to secondary characteristic classes. The latter turn out to be obstructions to the transversality of the Lagrangian subbundles, and we call them *Maslov classes* since the simplest of them is precisely the one that yields the Maslov index (See references in Section 1.1.)

4.1. CONNECTIONS ON PRINCIPAL BUNDLES.

There are various approaches to the subject of secondary characteristic classes [38], [15], [11], [43], [49], etc., and connections on principal bundles play an essential role in this subject. This is particularly true for the approach which we choose here and which is due to Chern-Simons [18], Bott [15], and D. Lehmann [49]. The reader is assumed to be familiar with the theory of connections on principal bundles (see [12], [45]). The aim of the present section is only to remind some necessary preparatory results that involve such connections.

Let $\pi : P \to M$ be a differentiable principal bundle whose structure group is the Lie group G with the corresponding Lie algebra $\mathfrak{g}$. Then a G-*connection* on P consists of a *connection form* θ, which is a differentiable $\mathfrak{g}$-valued 1-form on P such that: i) if v is tangent to the fiber of P at $p \in P$, then $\theta(v) = X \in \mathfrak{g}$ where X generates v (i.e., v is the velocity at p of the path obtained by acting with exp (tX) at the right on p); ii) for every $\gamma \in G$ if R_γ denotes the right translation of P by γ, one has $R^*_\gamma\theta = (\mathrm{ad}\,\gamma^{-1})\,\theta$.

Usually, we shall denote connection forms by $\theta, \theta', \theta_1, \ldots$. If θ is a connection form, $\theta = 0$ defines the so-called *horizontal distribution* H of the connection which is such that: 1) $H \oplus V = TP$ where V is the *vertical distribution* of the spaces tangent to the fibers of P; 2) H is invariant by all $R_{\gamma *}$ $(\gamma \in G)$. The definition of H satisfying 1), 2) is equivalent to the definition of the connection. If $v \in TP$, it has a unique decomposition $v = v_H + v_V$ into a *horizontal* and a *vertical part*.

Furthermore, let $\{U_\alpha\}_{\alpha \in A}$ be a covering of M by coordinate neighbourhoods which are trivializing for P, i.e., such that there are diffeomorphisms

$$(4.1.1) \qquad h_\alpha : \pi^{-1}(U_\alpha) \to U_\alpha \times G\ .$$

Let σ_α be local cross sections of P along U_α. Then, we must have relations

$$(4.1.2) \qquad \sigma_\beta(u) = R_{\gamma_{\alpha\beta}(u)}\sigma_\alpha(u) \qquad (u \in U_\alpha \cap U_\beta,\ \gamma_{\alpha\beta} \in G)\ .$$

(Such local sections exist because of (4.1.1). For instance, we may take $\sigma_\alpha(u) = h_\alpha^{-1}(u,e)$, where $u \in U_\alpha$, and e is the unit of G. In this case $\gamma_{\alpha\beta}$ are the transition functions corresponding to (4.1.1). Conversely, any system $\{\sigma_\alpha\}$ has an associated system of trivialization $\bar{h}_\alpha$ such that $\sigma_\alpha(u) = \bar{h}_\alpha(u,e)$.) A connection form θ on P induces the system of local $\mathfrak{g}$-valued 1-forms defined on U_α

$$(4.1.3) \qquad \theta_\alpha = \sigma^*_\alpha\theta\ ,$$

and (4.1.2) implies [45]

(4.1.4) $$\theta_\beta = (\mathrm{ad}\ \gamma_{\alpha\beta}^{-1})\theta_\alpha + \gamma_{\alpha\beta}^{-1}\, d\gamma_{\alpha\beta}$$

over $U_\alpha \cap U_\beta$. ($d\gamma_{\alpha\beta}$ is the derivative of the mapping $\gamma_{\alpha\beta}$.) Conversely, it is possible to reconstruct θ from the system $\{\theta_\alpha\}$, and the result is that

(4.1.5) $$\theta\big|_{\pi^{-1}(U_\alpha)} = (\mathrm{ad}\ \gamma^{-1})\theta_\alpha + \gamma^{-1}\, d\gamma \qquad (\gamma \in G)$$

is the expression of the connection form via trivializations (4.1.1) [45].

If θ_a $(a = 1,2)$ are connections, and $\alpha = \theta_2 - \theta_1$, then α is a $\mathfrak{g}$-valued 1-form on P which vanishes on vertical arguments, and is such that $R_\gamma^*\alpha = (\mathrm{ad}\ \gamma^{-1})\alpha$. We say that α is a *tensorial 1-form of type ad*, and one defines similarly *tensorial* k-*forms, of type ad* as $\mathfrak{g}$-valued k-forms φ on P such that $\varphi = 0$ if at least one of the arguments is vertical, and $R_\gamma^*\varphi = (\mathrm{ad}\ \gamma^{-1})\varphi$. For such forms, we may again define corresponding local systems $\varphi_\alpha = \sigma_\alpha^*\varphi$ like in (4.1.3), and then we have [45]

(4.1.6) $$\varphi_\beta = (\mathrm{ad}\ \gamma_{\alpha\beta}^{-1})\varphi_\alpha \quad .$$

The reconstruction of φ from the φ_α is by

(4.1.7) $$\varphi\big|_{\pi^{-1}(U_\alpha)} = \mathrm{ad}\ \gamma^{-1}\varphi_\alpha \qquad (\gamma \in G).$$

Proposition 4.1.1. *a) If* θ *is a connection on* P *all the other connections of* P *are given by* $\theta+\alpha$ *where* α *is an arbitrary* $\mathfrak{g}$*-valued tensorial 1-form of type ad. b) The set of all the connections on* P *is an (infinite dimensional) affine space, hence a convex linear combination of connection forms is a connection form.*

Proof. a) is obvious, and $\{\alpha\}$ is a (infinite dimensional) vector space

A. If C is the set of connections, then $\rho : C \times C \to A$ given by $\rho(\theta_1,\theta_2) = \theta_2 - \theta_1$ provides us with the requested affine structure of C. (See the proof of Theorem 2.2.15.) Finally, remember that a convex combination of $\theta_0,\theta_1,\ldots,\theta_r$ is

$$(4.1.8) \qquad \tilde{\theta} = \sum_{h=0}^{r} t_h \theta_h \quad \left(\sum_{h=0}^{r} t_h = 1,\ t_h \geq 0\right) \;,$$

whence we see that $\tilde{\theta} \in C$. Q.e.d.

Now, let again P be a G-principal bundle with a connection θ having the horizontal distribution H, and let λ be a Q-valued k-form where Q is an arbitrary vector space. Then, the formula

$$(4.1.9) \qquad \nabla\lambda(v_0,v_1,\ldots,v_q) = d\lambda(v_{0_H},v_{1_H},\ldots,v_{q_H})$$

defines the θ-*covariant derivative* of λ. It is clear that $\nabla\lambda$ vanishes if at least one argument is vertical. Particularly the form

$$(4.1.10) \qquad \Theta = \nabla\theta$$

is called the *curvature form* of the connection, and *it is a g-valued tensorial 2-form of type ad.* (Generally, if $\theta,\theta',\ldots$ are connection forms, we denote by $\Theta,\Theta',\ldots,$ the corresponding curvature forms.) Checking for horizontal and vertical arguments yields the following expression for the curvature form [12, 45].

$$(4.1.11) \qquad \Theta = d\theta + [\theta,\theta]$$

where the bracket denotes the product in the Lie algebra g and

$$(4.1.12) \qquad [\theta,\theta](v,w) = [\theta(v),\theta(w)] \;.$$

Another very important property of the curvature is the *Bianchi identity*

$$(4.1.13) \qquad \nabla\Theta = 0 \;.$$

We are also interested to quote the following formula of the covariant derivative of a tensorial 1-form λ of type ad [45],

(4.1.14) $$\nabla\lambda = d\lambda + [\lambda,\theta] + [\theta,\lambda] \quad .$$

$\nabla\lambda$ is a tensorial 2-form of type ad.

Finally, let us recall that if the principal bundle P has a principal subbundle P' with the structure group $G' \subset G$ (i.e., the structure group G of P can be reduced to G'), and if θ' is a connection on P', then θ' extends to a unique connection θ on P. This can be achieved by acting on the horizontal spaces of θ' with R_γ for all $\gamma \in G$. Or, else, if θ'_α are the local connection forms of θ', θ is defined by (4.1.5) where we put θ'_α instead of θ_α. A connection θ which extends a connection on P' will be called a G'-*connection on* P.

The applications that we have in mind are for principal bundles of linear frames (bases) tangent to a given manifold M^n. If $\mathcal{B}(M)$ denotes the set of all the bases of the fibers of TM, then $\pi : \mathcal{B}(M) \to M$ (the natural projection) is a $G\ell(n,\mathbb{R})$-principal bundle. The local cross sections σ_α of the general case are now local fields of tangent bases $(e_i^{(\alpha)})$ $(i = 1,\dots,n)$, and the transition relations (4.1.2) can be written as

(4.1.15) $$e_i^{(\beta)} = \underset{(\alpha,\beta)}{\xi_i^h} \, e_h^{(\alpha)} \qquad ,$$

where $\underset{(\alpha,\beta)}{\xi} = \left(\underset{(\alpha,\beta)}{\xi_i^h} \right) \in G\ell(n,\mathbb{R})$, and it depends differentiably on the base point in $U_\alpha \cap U_\beta$. (Remember that we denote entries of a matrix with the upper index as a line index, and with the lower index as a column index, i.e., the matrix form of (4.1.15) is $e^{(\beta)} = e^{(\alpha)}\xi$.)

Accordingly, a connection θ on $\mathcal{B}(M)$, which is called a *linear connection* on M, is defined by the local connection forms θ_α which are now matrices of scalar 1-forms over the neighbourhoods U_α namely $\theta_\alpha = (\underset{(\alpha)}{\theta^j_i})$, and it is usual to write down the so-called *local equations of the connection* θ as

$$(4.1.16)\qquad \nabla e_i^{(\alpha)} = \underset{(\alpha)}{\theta^j_i}\, e_j^{(\alpha)} \qquad\qquad (\nabla e^{(\alpha)} = e^{(\alpha)}\theta)\ .$$

By means of coordinates (x^i) on U_α, one has $\underset{(\alpha)}{\theta^j_i} = \underset{(\alpha)}{\Gamma^j_{ih}}\, dx^h$, and Γ^j_{ih} are the *connection coefficients.*

The operator ∇ of (4.1.16) can be seen as a "directional derivative" ∇_X where $X \in TM$, and it induces the usual covariant derivative of tensor fields [45]. The equations (4.1.4) become now

$$(4.1.17)\qquad \underset{(\beta)}{\theta} = \xi^{-1}\ \underset{(\alpha)}{\theta}\ \xi + \xi^{-1} d\xi \quad ,$$

and if $\eta = \xi^{-1}$ this means

$$(4.1.18)\qquad \underset{(\beta)}{\theta^j_i} = \eta^j_h \xi^k_i\ \underset{(\alpha)}{\theta^h_k} + \eta^j_h d\xi^h_i \quad .$$

Similarly, if (x^i) are local coordinates on M and (ξ^j_i) are coordinates on $G\ell(n,\mathbb{R})$ (entries of matrices), the global connection form θ is given by (4.1.5), i.e.,

$$(4.1.19)\qquad \theta^j_i\Big|_{\pi^{-1}(U_\alpha)} = \eta^j_h \xi^k_i\ \underset{(\alpha)}{\theta^h_k} + \eta^j_h d\xi^h_i \quad .$$

Particularly, if θ is a G'-connection on $\mathcal{B}(M)$ for $G' \subset G\ell(n,\mathbb{R})$, formula (4.1.19) offers us the expression of the connection form on $\mathcal{B}(M)$, if ξ runs over the whole group $G\ell(n,\mathbb{R})$.

Furthermore, the curvature of a linear connection is a matrix of scalar 2-forms $\Theta = (\Theta^j_i)$, and since it is a tensorial form of type ad, we may write by (4.1.7) that

$$(4.1.20)\qquad \Theta^j_i\Big|_{\pi^{-1}(U_\alpha)} = \eta^j_h \xi^k_i \underset{(\alpha)}{\Theta}{}^h_k ,$$

where $\underset{(\alpha)}{\Theta}{}^h_k$ are the local curvature forms. (These are scalar 2-forms on U_α.) By using (4.1.11), one obtains

$$(4.1.21)\qquad \underset{(\alpha)}{\Theta}{}^h_k = d\,\underset{(\alpha)}{\theta}{}^h_k - \underset{(\alpha)}{\theta}{}^j_k \wedge \underset{(\alpha)}{\theta}{}^h_j .$$

Similarly, if λ is a tensorial 1-form of type ad, it is determined by local matrices of 1-forms $(\underset{(\alpha)}{\lambda}{}^j_i)$, and its covariant derivative (4.1.14) is given by the local matrices

$$(4.1.22)\qquad (\underset{(\alpha)}{\nabla\lambda})^j_i = d\,\underset{(\alpha)}{\lambda}{}^j_i + \underset{(\alpha)}{\theta}{}^j_h \wedge \underset{(\alpha)}{\lambda}{}^h_i - \underset{(\alpha)}{\theta}{}^h_i \wedge \underset{(\alpha)}{\lambda}{}^j_h .$$

The main subgroup of $G\ell(n,\mathbb{R})$ which we shall need later is the orthogonal group $O(n)$. A reduction of the structure group of $\mathcal{B}(M)$ to $O(n)$ is equivalent to a riemannian metric g on M, and the corresponding subbundle is the bundle of tangent g-orthonormal frames of M. Since the Lie algebra of $O(n)$ consists of skew symmetric matrices, we have the following fundamental facts

Proposition 4.1.2. *If (4.1.16) are the local equations of a* $O(n)$*-connection on* M, *and if* $(e^{(\alpha)}_i)$ *are orthonormal frames then the connection forms, and the curvature forms satisfy*

$$(4.1.23)\qquad \underset{(\alpha)}{\theta}{}^j_i + \underset{(\alpha)}{\theta}{}^i_j = 0 \;;\quad \underset{(\alpha)}{\Theta}{}^j_i + \underset{(\alpha)}{\Theta}{}^i_j = 0 .$$

Finally, let us remember that g admits a canonical unique connection which is an $O(n)$-connection without *torsion*. This is the classical *Levi-Civita connection*, and the corresponding connection coefficients Γ^k_{ij} are the *Christoffel symbols* of g [45, 31].

Now, let us come back to general G-connections on principal fibre bundles. Later on, we shall make some reference to a classification theorem of such connections, and we shall explain this theorem here. In topology, it is known that for any topological group G and for any natural number k one can construct a k-*classifying space* B_G^k, defined up to a homotopy equivalence and endowed with a k-*universal principal* G-*bundle* $E_G^k \to B_G^k$ such that for any principal G-bundle $P \to X$ over a CW-complex X with $\dim X < k$ one has $P = \varphi^* E_G^k$ where $\varphi : X \to B_G$ is the *classifying map* that exists, and it is defined up to a homotopy. Particularly, for $G = O(n)$, one can take for B_G^k a *Grassmann manifold* of n-planes $\mathfrak{G}(m,n)$ with $m \geq k+n$, and for E_G^k the corresponding *Stiefel manifold* $\mathcal{V}(m,n)$ of n-orthonormal frames in $\mathbb{R}^m$ with its natural fibration over $\mathfrak{G}(m,n)$. Similarly, for $G = U(n)$, one has the same result but for the complex Grassmann and Stiefel manifolds. It is also usual to define ∞-classifying spaces and universal bundles by taking inductive limits on k. These universal bundles are denoted by $E_G \to B_G$, and we have $P = \varphi^* E_G$ for any G-principal bundle over a CW-complex and for a corresponding classifying map $\varphi : M \to B_G$, defined up to homotopy [29, 42, 76].

Now, the objects of our interest are differentiable principal G-bundles $\pi : P \to M$ endowed with a connection θ, where G is a Lie group. It is easy to see that if $\varphi : P_1 \to P_2$ is a principal bundle mapping (i.e., it commutes with the right action of G on the two spaces), and θ_2 is a connection on P_2, then $\theta_1 = \varphi^*\theta_2$ is a connection on P_1. We may use this fact in order to state that the objects $(P \to M, \theta)$ form a *category* P_G whose morphisms are *connection compatible bundle mappings*. (Hence a bundle mapping $\varphi : P_1 \to P_2$ is a morphism in P_G iff $\theta_1 = \varphi^*\theta_2$.) In this situation, a classification theorem has been proven by Narasimhan and Ramanan [68], and (if combined with the topological results) it states

Theorem 4.1.3. [68]. *Let* G *be a Lie group with a finite number of connected components. Then, for any positive integer* k, *there exists a differentiable principal* G-*bundle* $E_G^k \to B_G^k$ *with a connection* $\theta_G^{(k)}$ *such*

that for any object $(P \to M, \theta)$ *of* P_G *with* $\dim M < k$ *there is mapping* $f : M \to B_G^k$ *defined up to homotopy such that* P *is the pull back of* E_G^k *by* f*, and, if* $\varphi : P \to E_G^k$ *is the corresponding lift of* f*, one has* $\theta = \varphi^* \theta_G^{(k)}$.

Definition 4.1.4. *The bundle with connection* $\left(E_G^k \to B_G^k, \theta_G^{(k)}\right)$ *of Theorem 4.1.3 is called a* ***k-universal object of*** P_G*, and* $\theta_G^{(k)}$ *is a* ***k-universal*** ***G-connection.*** *The mapping* f *is a classification mapping.*

Remark 4.1.5. *If we take the inductive limit for* $k \to \infty$*, we do not get differentiable manifolds. However, we shall use the notation* $(E_G \to B_G, \theta_G)$*, which will refer simply to the objects of Definition 4.1.4 for an arbitrarily large* k.

The idea of the proof of Theorem 4.1.3 is as follows. The result is proven first for $G = O(n)$, $B_G^k = \mathfrak{G}(m,n)$, $E_G^k = \mathfrak{V}^P(m,n)$, and the natural connection θ_0 on this bundle. For a compact group G, the result is obtained by using an orthogonal representation of G, and for a connected group it is obtained by an adequate use of a reduction to a maximal compact subgroup K. In [68] it is also proven that in the case $G = U(n)$, B_G^k and E_G^k are the corresponding *Grassmann and Stiefel manifolds.*

We shall close this preparatory section by recalling a subject not related to connections namely, the *Stokes formula* for integration along the fibers of a (not necessarily principal) locally trivial fiber bundle $\pi : W^w \to M^m$ of fibre F^f. The fibre F may have a boundary ∂F, and even edges and corners included in ∂F, and then differentiability (which we always assume) is understood by extending F to a smooth open neighbourhood.

Let us consider the case where $W = M \times F$, while F has an orientation defined by an f-form κ, and ∂F has the induced orientation

$$\xi(e_1, \ldots, e_{n-1}) = \kappa(e_{ext}, e_1, \ldots, e_{n-1}) \tag{4.1.24}$$

where $e_1,\ldots,e_{n-1}$ is a basis of $T\partial F$, and e_{ext} is tangent to M at the same point of ∂F, and it "looks outward." Then, we say that κ defines a *bundle orientation* of W.

Now, we can define a linear mapping

$$\int_F : \Lambda_c^k W \to \Lambda^{k-f} M \tag{4.1.25}$$

of the real vector space $\Lambda_c^k W$ of the k-forms on W with compact support along the fibers to the space of the (k-f)-forms of M which is called *integration along the fibers*. Namely, if $\alpha \in \Lambda_c^k W$ but α does not belong to the ideal (κ) generated by the bundle orientation of W, we set $\int_F \alpha = 0$. For $\alpha \in (\kappa)$, we have

$$\alpha = \lambda \wedge \kappa \quad , \tag{4.1.26}$$

where if (x_i) are local coordinates about $x \in M$, and $t \in F$ then

$$\lambda = \frac{1}{(k-f)!}\lambda_{i_1\ldots i_{k-f}}(x^j,t)\,dx^{i_1} \wedge \ldots \wedge dx^{i_{k-f}} \quad . \tag{4.1.27}$$

Then, we shall define

$$\int_F \alpha = \frac{1}{(k-f)!}\,dx^{i_1} \wedge \ldots \wedge dx^{i_{k-f}} \cdot \left(\int_F \lambda_{i_1\ldots i_{k-f}}(x^j,t)\kappa\right) \tag{4.1.28}$$

where the integral at the end is usual integration of a form on a manifold. Clearly, the form (4.1.28) does not depend on the choice of the local coordinates (x^i).

In the case of a general fibration $\pi : W \to M$ of fibre F, we shall assume that the bundle orientation exists, i.e., κ induces orientations in the local fibers $F_x = \pi^{-1}(x)$ $(x \in M)$ via the local trivializations of W, and there is a covering $\{U_\alpha\}_{\alpha \in A}$ of M by neighbourhoods such that $W/_{U_\alpha}$ is trivial, and for $x \in U_\alpha \cap U_\beta$ the induced orientations in F_x are equal. Then we shall define $\oint_F \alpha$ over each U_α, and the results glue up to a global form. This provides us with the operation of integration along fibers for the general fiber bundle W. (See details in [16] and, even more, in [34].)

The basic fact about fiber integration which we shall need is [34], [15]

Theorem 4.1.6. (Stokes' formula). *Let* $\pi : W \to M$ *be a locally trivial fiber bundle of fibre* F^f *which admits a bundle orientation defined by an orientation* κ *of* F. *Let* ∂F *be the boundary of* F *with induced orientation. Then* $\pi : \partial W \to M$ *is a fiber bundle with fiber* ∂F, *and it has an induced bundle orientation. Let* $\iota : \partial W \to W$ *be the natural inclusion. Then, for every form* $\alpha \in \Lambda_c^k W$ *one has*

$$\int_F d\alpha - d\int_F \alpha = (-1)^{k-f+1}\int_{\partial F} \iota^*\alpha \quad . \tag{4.1.29}$$

Proof. (Sketch). Using local trivializations, we see that it is enough to prove (4.1.29) for $W = M \times F$. If $\alpha \notin (\kappa)$ and also $d\alpha \notin \kappa$ the equality (4.1.29) is trivial. If $\alpha \notin (\kappa)$ but $d\alpha \in (\kappa)$, (4.1.29) is nothing else but the usual Stokes' formula on the manifold F [59]. (The sign $(-1)^{k-f+1}$ appears because of the order of the factors in the expression of the type (4.1.26) for $d\alpha$.) Finally, if $\alpha \in (\kappa)$, and it is given by (4.1.26), then $\iota^*\alpha = \lambda \wedge (\iota^*\kappa) = 0$, while, on the other hand $d\alpha = d\lambda \wedge \kappa$, and formula (4.1.28) yields straightforwardly $\oint_F d\alpha = d\oint_F \alpha$. Hence (4.1.29) is true again. Q.e.d.

4.2. SECONDARY CHARACTERISTIC CLASSES. This section is devoted to the general theory of *secondary characteristic classes* following the approach of Chern-Simons [18], Bott [15] and D. Lehmann [49]. Other approaches to this subject can be found in [38], [11] and [43].

In the theory of fiber bundles, one is led naturally to the

primary characteristic classes (see, for instance, [60] or [42]) of the principal bundles with structure group G. Namely, as we saw in Section 4.1, a G-principal bundle $P \to M$ is induced by a classifying mapping $f : M \to B_G$ defined up to homotopy. f induces a mapping in cohomology $f^* : H^*(B_G) \to H^*(M)$ and $\operatorname{im} f^*$ is the set of the primary characteristic classes of P. In this way, the primary characteristic classes allow us to distinguish between various G-principal bundles.

In the present work, we are interested only in real valued characteristic classes. The real cohomology algebra $H^*(B_G, \mathbb{R})$ has been computed for compact connected Lie groups G, and we call it the *invariant Weil algebra* of G (see [43], for example, for the use of the term *Weil algebra)*. Conversely, there exists a procedure known as the *Chern-Weil homomorphism* which yields the characteristic classes directly from the Weil algebra, and which may also be used in the noncompact case. Moreover, this procedure leads also to the secondary characteristic classes. It is for this reason that we shall start with the Weil algebra, as the basic universal object in the construction of characteristic classes.

Let G be a Lie group with the Lie algebra g. The main objects to be considered are the symmetric multilinear real valued functions

$$(4.2.1) \qquad f : \underbrace{g \times \ldots \times g}_{k \text{ times}} \to \mathbb{R}$$

that are ad γ-invariant, i.e.,

$$(4.2.2) \qquad f\big(\operatorname{ad}\gamma(v_1), \ldots, \operatorname{ad}\gamma(v_k)\big) = f(v_1, \ldots, v_k)$$

for every set $v_\alpha \in g$ $(\alpha = 1, \ldots, k)$, and $\gamma \in G$.

Definition 4.2.1. *A function* f *as in (4.2.1), which satisfies (4.2.2) is called an* ***invariant symmetric multilinear function of grade*** $2k$ *on* g, *and the* $\mathbb{R}$*-linear space of these functions is denoted by* $I^k(G)$. *The space*

$$(4.2.3)\qquad I(G) = \bigoplus_{k\geq 0} I^k(G)$$

has the natural structure of an $\mathbb{R}$*-algebra whose product is the* ***symmetric product*** *of functions* f, *and this algebra is called the* ***invariant Weil algebra*** *of* G.

Any function (4.2.1) gives rise to a *polynomial function* f : $\bar{f} : g \to \mathbb{R}$, i.e., a function $\bar{f}(v)$ $(v \in g)$ that is a homogeneous polynomial of degree k in the coordinates of v with respect to a basis of g. Namely, one sets

$$(4.2.4)\qquad \bar{f}(v) = f(\underbrace{v,\dots,v}_{k \text{ times}}) .$$

Conversely, the polynomial $\bar{f}$ defines f by "polarization", i.e., if $v = \xi^i e_i$, where e_i is a basis of g, and if

$$(4.2.5)\qquad \bar{f}(v) = \frac{1}{k!}\alpha_{i_1\dots i_k}\xi^{i_1}\dots\xi^{i_k} \qquad ,$$

where $\alpha\dots$ are symmetric coefficients, then

$$(4.2.6)\qquad f(v_1,\dots,v_k) = \frac{1}{k!}\alpha_{i_1\dots i_k}\xi^{i_1}_{(1}\dots\xi^{i_k}_{k)}$$

where $v_\lambda = \xi^i_\lambda e_i$ $(\lambda = 1,\dots,k)$, and the product of the coordinates ξ is symmetrized in (4.2.6). This procedure does not depend on the choice of the basis (e_i).

It is clear that f satisfies (4.2.2) iff $\bar{f}$ satisfies the same invariance property. Accordingly, I(G) is isomorphic to the similarly constructed algegra of ad G-invariant polynomials. In the sequel, we shall identify f and $\bar{f}$, and talk indistinctly of the elements of I(G) as either invariant functions or invariant polynomials. We shall also use the notation

$$(4.2.7)\qquad f\left(v_1^{(k_1)},v_2^{(k_2)},\dots,v_s^{(k_s)}\right) = f(\underbrace{v_1,\dots,v_1}_{k_1 \text{ times}},\underbrace{v_2,\dots,v_2}_{k_2 \text{ times}},\dots,\underbrace{v_s,\dots,v_s}_{k_s \text{ times}}),$$

where $f \in I^k(G)$; $v_1,\dots,v_s \in \mathcal{g}$; $k_1 + \dots + k_s = k$.

Finally, we shall introduce differential forms obtained from (4.2.6) by using $\mathcal{g}$-valued differential forms as arguments v, and the exterior product of the component forms ξ in the right-hand side of formula (4.2.6).

(The reader who would like to have more details about the previous definitions can find them, for instance, in [45].)

The characteristic classes of a principal bundle $P \to M$ are produced by evaluating invariant polynomials on arguments which are curvature forms of connections on P. We shall start by establishing a general formula of R. Bott [15].

Let $\pi : P \to M$ be a G-principal bundle, and let θ_h $(h = 0,1,\dots,r)$ be $r+1$ connection forms on P. Let Δ^r denote the standard r-symplex

$$\Delta^r = \{(t_0,t_1,\dots,t_r) \in \mathbb{R}^{r+1} / t_h \geq 0, \sum_{h=0}^{r} t_h = 1\} . \tag{4.2.8}$$

Then we have a corresponding G-principal bundle

$$\pi \times id : P \times \Delta^r \to M \times \Delta^r \tag{4.2.9}$$

(which can be "smoothed" at the boundary), and, in view of Proposition 4.1.1.b), the form

$$\tilde{\theta} = \sum_{h=0}^{r} t_h \theta_h \tag{4.2.10}$$

is a connection on $P \times \Delta^r$ which we call the *average connection*. (For every $t \in \Delta^r$, $\tilde{\theta}$ is also the *average connection* on P.) Now, let $f \in I^k(G)$, and let us define the form

$$\Delta(\theta_0,\dots,\theta_r)f = (-1)^{[\frac{r+1}{2}]} \int_{\Delta^r} f(\tilde{\Theta}^{(k)}) \tag{4.2.11}$$

where $\tilde{\Theta}$ is the curvature of the average connection, and the integration is along the fibers Δ^r of the projection $P \times \Delta^r \to P$ with the orientation $dt_1 \wedge \ldots \wedge dt_r$ $(t_0 = 1 - \sum_{k=1}^{r} t_k)$. Moreover, if we use for $\tilde{\theta}$ its local expression of the form (4.1.7), and since f satisfies (4.2.2), we see that $\Delta(\theta_0,\ldots,\theta_r)f$ may be computed by using the local curvature forms of $\tilde{\theta}$ on M instead of $\tilde{\Theta}$, and the result projects to a form on M. (In the sequel, if no confusion is feared, we identify forms on M with their lift to P.) Accordingly, we state

Definition 4.2.2. *The mappings*

$$(4.2.12) \qquad \Delta(\theta_0,\ldots,\theta_r) : I^k(G) \to \Lambda^{2k-r}(M)$$

are called Bott homomorphisms.

The basic property of the mappings (4.2.12) is [15]

Proposition 4.2.3. *For every* $f \in I^k(G)$, *and connections* $\theta_0,\ldots,\theta_r$ $(r \geq 0)$ *one has*

$$(4.2.13) \qquad d\Delta(\theta_0,\ldots,\theta_r)f = \sum_{h=0}^{r} (-1)^h \Delta(\theta_0,\ldots,\theta_{h-1},\theta_{h+1},\ldots\theta_r)f \; .$$

Proof. First, let us take $r=0$, and prove that $\Delta(\theta_0)f$ are closed forms. Indeed, we have

$$(4.2.14) \qquad \Delta(\theta_0)f = f\left(\Theta_0^{(k)}\right)$$

and this is the lift to P of a form on M. Hence this form, and its differential vanish if at least one of their arguments is a vertical vector, and (4.1.9) yields the covariant derivative

$$\nabla\left(f\left(\Theta_0^{(k)}\right)\right) = d\left(f\left(\Theta_0^{(k)}\right)\right) \qquad .$$

But it is clear that ∇ behaves in computations like d. Hence we get

$$(4.2.15) \qquad d\left(\Delta(\theta_0)f\right) = \sum_{j=1}^{k} f\left(\Theta_0,\ldots,\nabla\Theta_0,\ldots,\Theta_0\right) \; .$$

Because of the Bianchi identity (4.1.13), we finally get

(4.2.16) $$d\{\Delta(\theta_0)f\} = 0 .$$

Particularly, $\alpha = f(\tilde{\theta}^{(k)})$ is a closed form, and by Stokes' formula (4.1.29) we get

$$d\Delta(\theta_0,\ldots,\theta_r)f = (-1)^{[\frac{r+1}{2}]+2k-r}\int_{\partial\Delta^r} \iota^*\alpha =$$

$$= (-1)^{[\frac{r+1}{2}]+2k-r}\int_{\sum_{h=0}^{r}(-1)^h\Delta^r_{(h)}} \iota^*\alpha =$$

$$= (-1)^{[\frac{r+1}{2}]+[\frac{r}{2}]+2k-r}\sum_{h=0}^{r}(-1)^h\Delta(\theta_0,\ldots,\theta_{h-1},\theta_{h+1},\ldots\theta_r)f .$$

Since $\left[\frac{r+1}{2}\right]+\left[\frac{r}{2}\right] = r$, this is precisely the requested formula (4.2.13). Q.e.d.

Remark 4.2.4. *For* $r=0$, *the Bott homomorphism reduces to (4.2.14), and it is in fact an algebra homomorphism*

(4.2.17) $$\Delta(\theta_0): I(G) \to \Lambda(M) .$$

$\Delta(\theta_0)$ *is called the* ***Chern-Weil homomorphism****, and formula (4.2.16) has been known since the beginning of the Chern-Weil theory of characteristic classes. So was the particular case* $r=1$ *of (4.2.13) which is*

(4.2.18) $$d\Delta(\theta_0,\theta_1)f = \Delta(\theta_1)f - \Delta(\theta_0)f .$$

The results of Remark 4.2.4 justify

Definition 4.2.5. *Let us denote by brackets the real cohomology classes defined by the closed forms of* M *via de Rham's theorem (e.g., [80]).*

Then, the induced mapping

$$\text{(4.2.19)} \qquad w(G,P) = [\Delta(\theta_0)]: I(G) \to H^*(M,\mathbb{R})$$

is a homomorphism of graded algebras called the ***cohomological Chern-Weil homomorphism.*** *The cohomology classes of* im $w(G,P)$ *are called the* ***primary characteristic classes*** *of* P. *Every polynomial* $f \in I^k(G)$ *defines the class* $[\Delta(\theta_0)f] \in H^{2k}(M,\mathbb{R})$.

It follows from (4.2.18) that $w(G,P)$ does not depend on the choice of the connection since the forms $\Delta(\theta_0)f$ and $\Delta(\theta_1)f$ are cohomologous. Hence, indeed, *the primary characteristic classes depend only on* P. Then, if

$$\text{(4.2.20)} \qquad \begin{array}{ccc} P & \xrightarrow{\tilde{\varphi}} & P' \\ \downarrow & & \downarrow \\ M & \xrightarrow{\varphi} & M' \end{array}$$

is a bundle morphism we may use $\tilde{\varphi}$-related connection forms, and we get the commutative diagram

$$\text{(4.2.21)} \qquad \begin{array}{ccc} & \overset{w(G,P)}{\nearrow} & H^*(M,\mathbb{R}) \\ I(G) & & \uparrow \varphi^* \\ & \underset{w(G,P')}{\searrow} & H^*(M',\mathbb{R}) \end{array} .$$

Therefore, *the characteristic classes have a functorial character* [80]. Starting with this remark, one can proceed by an inductive limit, and get a *universal Chern-Weil homomorphism* $w(G)$ such that

$$\text{(4.2.22)} \qquad \begin{array}{ccc} & \overset{w(G)}{\nearrow} & H^*(B_G,\mathbb{R}) \\ I(G) & & \downarrow \varphi^* \\ & \underset{w(G,P)}{\searrow} & H^*(M,\mathbb{R}) \end{array}$$

is a commutative diagram, where $\varphi: M \to B_G$ is a classifying mapping of P. As mentioned earlier, if G is compact, $w(G)$ is an isomorphism [14,42,60], and this proves the relation between Definition 4.2.5 and the usual topological definition of characteristic classes.

Furthermore, formula (4.2.18) may be used for another purpose too. Namely, it shows that if θ_0 and θ_1 are fixed, and if $f \in \ker \Delta(\theta_0) \cap \ker \Delta(\theta_1)$, $\Delta(\theta_0, \theta_1)f$ is a closed form. We give

Definition 4.2.6. *If* $f \in I^k(G)$ *belongs to* $\ker \Delta(\theta_0) \cap \ker \Delta(\theta_1)$, *then the cohomology class*

$$[\Delta(\theta_0, \theta_1)f] \in H^{2k-1}(M, \mathbb{R}) \tag{4.2.23}$$

*is called a **simple secondary characteristic class of the triple*** (P, θ_0, θ_1).*)

Since the differential forms $\Delta(\theta_0, \theta_1)f$ are fundamental, let us describe them also as follows. These forms are defined by (4.2.11) with r=1. Let us denote the coordinates in Δ^1 by $t_0 = 1-t$, $t_1 = t \in [0,1]$. then (4.2.10) gives

$$\tilde{\theta} = \theta_0 + t\alpha \quad , \quad \alpha = \theta_1 - \theta_0 \ , \tag{4.2.24}$$

and (4.1.11) yields for the corresponding curvature form

$$\tilde{\Theta} = dt \wedge \alpha + \Theta_t \ , \tag{4.2.25}$$

where Θ_t is the curvature of $\tilde{\theta}|_{t=const} \overset{def}{=} \theta_t$ on P. By using this expression of $\tilde{\Theta}$, formula (4.2.11) becomes

$$\Delta(\theta_0, \theta_1)f = k \int_0^1 f(\alpha, \Theta_t^{(k-1)})dt \qquad (f \in I^k(G)). \tag{4.2.26}$$

Before we proceed with the study of the secondary classes, it is

---

*) A class (4.2.23) is also obtained if $f \in \ker(\Delta(\theta_1) - \Delta(\theta_0))$ but these classes are less interesting for applications.

interesting to construct similar differential forms on the total space P of the G-principal bundle $\pi\colon P \to M$ [18].

Let θ be a connection on P, and let $t\theta$ $(0 \leqslant t \leqslant 1)$ be a corresponding form on $P \times \Delta^1$. While $t\theta$ is not a connection, we may, however, associate with it a "curvature"

$$(4.2.27) \qquad \overline{\Theta} = d(t\theta) + [t\theta,t\theta] = dt \wedge \theta + \overline{\Theta}_t \quad ,$$

where

$$(4.2.28) \qquad \overline{\Theta}_t = td\theta + t^2[\theta,\theta] = t\Theta - t(1-t)[\theta,\theta] = t\Theta + t(1+t)d\theta \quad ,$$

and Θ is the curvature form of θ.

Definition 4.2.7. *The differential forms*

$$(4.2.29) \qquad T(\theta)f = -\int_{\Delta^1} f\left(\overline{\Theta}^{(k)}\right) = k\int_0^1 f\left(\theta,\overline{\Theta}^{(k-1)}\right)dt$$

$\left(f \in I^k(G)\right)$ *are called* ***transgression forms.***

$T(\theta)f$ are $(2k-1)$-forms on P and, generally, they do not project to forms on M. They have the following important property [18]

Proposition 4.2.8. *For every* $f \in I^k(G)$ *one has*

$$(4.2.30) \qquad d\left(T(\theta)f\right) = \pi^*\Delta(\theta)f$$

where π *is the projection* $\pi\colon P \to M$.

Proof. It follows from (4.1.9) that $\nabla d\theta = 0$, and, since $\nabla\Theta = 0$ (4.2.28) implies $\nabla\overline{\Theta}_t = 0$ for all $0 \leqslant t \leqslant 1$. Hence, like in the proof of (4.2.13), $f\left(\overline{\Theta}^{(k)}\right)$ is a closed form. Now, if we apply the Stokes' formula (4.1.29) for $\alpha = f\left(\overline{\Theta}^{(k)}\right)$ on $P \times \Delta^1$, we get precisely (4.2.30). Q.e.d.

Accordingly, we may give [18]

Definition 4.2.9. *For every* $f \in I^k(G)$ *such that* $f \in \ker \Delta(\theta)$ *one has a cohomology class* $[T(\theta)f] \in H^{2k-1}(P, \mathbb{R})$, *and these classes are called the* ***Chern-Simons classes*** *of* (P,θ).

Remark 4.2.10. *The Chern-Simons classes are functorial in the category* P_G *of* G-*principal bundles with connections.*

Now, let us approach the question of the dependence of secondary and Chern-Simons classes on the connections. We shall do this by using a homotopy idea due to D. Lehmann [49]. Let us first start with

Definition 4.2.11. *Let* $I = \{s/0 \leq s \leq 1\}$, *let* φ *be a connection on the* G-*bundle* $P \times I \to M \times I$, *and let us denote by* $\theta_s = \varphi|_{s=\text{const}}$. *the corresponding family of connections on* P. *Then* φ *is called a* ***deformation*** *of* θ_0 *and, also, a* ***link*** *between* θ_0 *and* θ_1.

A connection φ like in Definition 4.2.11 is given by a form

(4.2.31) $$\varphi = \theta_s + \lambda ds \quad ,$$

where $\lambda: P \times I \to g$ is a function such that

(4.2.32) $$\lambda \circ R_\gamma = \mathrm{ad}\, \gamma^{-1} \circ \lambda \qquad (\gamma \in G).$$

The curvature form of φ can be computed by (4.1.11), and it is

(4.2.33) $$\Phi = \Theta_s + \left(d_P\lambda + [\theta_s, \lambda] - \frac{\partial \theta_s}{\partial s}\right) \wedge ds \quad ,$$

where Θ_s is the curvature of θ_s on P, and $d_P\lambda = d\lambda$ (modulo s=const.)

Proposition 4.2.12. *Let*

(4.2.34) $$\varphi = \theta_s + \lambda ds \quad , \quad \varphi' = \theta'_s + \lambda' ds$$

be two connections on $P \times I$. *Then, the following difference formulas hold good for every* $f \in I^k(G)$

(4.2.35) $$T(\theta_1)f - T(\theta_0)f =$$

$$= k\int_0^1 f\left(\frac{\partial\theta_s}{\partial s} - [\theta_s,\lambda] - d_P\lambda, \theta_s^{(k-1)}\right)ds + \text{exact form on } P =$$

$$= k\int_0^1 f\left(\frac{\partial\theta_s}{\partial s}, \theta_s^{(k-1)}\right) ds + \text{exact form on } P,$$

(4.2.36) $$\Delta(\theta_1,\theta_1')f - \Delta(\theta_0,\theta_0')f = k\int_0^1 \left\{ f\left(\frac{\partial\theta_s'}{\partial s} - [\theta_s',\lambda'] - d_P\lambda', \theta_s'^{(k-1)}\right) - \right.$$

$$\left. - f\left(\frac{\partial\theta_s}{\partial s} - [\theta_s,\lambda] - d_P\lambda, \theta_s^{(k-1)}\right)\right\} ds + \text{exact form on } M =$$

$$= k\int_0^1 \left\{ f\left(\frac{\partial\theta_s'}{\partial s}, \theta_s'^{(k-1)}\right) - f\left(\frac{\partial\theta_s}{\partial s}, \theta_s^{(k-1)}\right)\right\} ds +$$

$$+ \text{exact form on } M .$$

Proof. Let $i_s : P \to P\times I$ be defined by $i_s(p)=(p,s)$ $(p\in P, s\in I = [0,1])$. Then we have $i_s^*\big(T(\varphi)f\big) = T(\theta_s)f$. Now, if we apply the Stokes' formula (4.1.29) for $\alpha = T(\varphi)f$ on $P\times I$, and since $\deg\alpha = 2k-1$, we get

$$T(\theta_1)f - T(\theta_0)f = \int_I d\big(T(\varphi)f\big) - d\int_I T(\varphi)f ,$$

i.e., in view of (4.2.30)

(4.2.37) $$T(\theta_1)f - T(\theta_0)f = \int_I f\big(\Phi^{(k)}\big) + \text{exact form on } P.$$

But from (4.2.33) we obtain

(4.2.38) $$f\big(\Phi^{(k)}\big) = f\big(\theta_s^{(k)}\big) + kf\left(d_P\lambda + [\theta_s,\lambda] - \frac{\partial\theta_s}{\partial s}, \theta^{(k-1)}\right)\wedge ds .$$

Hence (4.2.37) is exactly the first equality (4.2.35).

Furthermore, let us take the form

$$\alpha = f\big(\lambda ds , \Phi^{(k-1)}\big)$$

on $P \times I$, and apply (4.1.29) to it. Since α is zero on ∂I, we get

$$\text{(4.2.39)} \qquad d\int_I f(\lambda ds, \Phi^{(k-1)}) = \int_I df(\lambda ds, \Phi^{(k-1)}) =$$

$$= \int_I f(\nabla_\varphi(\lambda ds), \Phi^{(k-1)}) = \int_0^1 f(d\lambda + [\theta_s, \lambda], \Theta_s^{(k-1)})ds ,$$

where ∇_φ is the covariant derivative with respect to φ (hence $\nabla_\varphi \Phi = 0$ because of the Bianchi identity), and the formula $\nabla_\varphi(\lambda ds) = \{d\lambda + [\theta_s, \lambda]\} \wedge ds$ can be checked straightforwardly by using (4.1.14)

If we use (4.2.39) in the already proven part of (4.2.35) we obtain the second part of (4.2.35).

Finally, formulas (4.2.36) can be obtained in exactly the same manner, by using Stokes' formula (4.1.29) for the form $\alpha = \Delta(\varphi, \varphi')f$ on M, by using (4.2.18), and by using the computation (4.2.39), which contains only projectable forms. Q.e.d.

Corollary 4.2.13. *Under the hypotheses of Proposition 4.2.12 we have the following derivation formulas*

$$\text{(4.2.40)} \qquad \frac{\partial(T(\theta_s)f)}{\partial s} = kf\left(\frac{\partial\theta_s}{\partial s} - [\theta_s, \lambda] - d_P\lambda , \Theta_s^{(k-1)}\right) +$$

$$+ \text{ exact form on } P = kf\left(\frac{\partial\theta_s}{\partial s}, \Theta_s^{(k-1)}\right) + \text{ exact form on } P,$$

$$\text{(4.2.41)} \qquad \frac{\partial(\Delta(\theta_s, \theta_s')f)}{\partial s} = k\left\{f\left(\frac{\partial\theta_s'}{\partial s} - [\theta_s', \lambda'] - d_P\lambda', \Theta_s'^{(k-1)}\right) - \right.$$

$$\left. - f\left(\frac{\partial\theta_s}{\partial s} - [\theta_s, \lambda] - d_P\lambda, \Theta_s^{(k-1)}\right)\right\} + \text{ exact form on } M =$$

$$= k\left\{f\left(\frac{\partial\theta_s'}{\partial s}, \Theta_s'^{(k-1)}\right) - f\left(\frac{\partial\theta_s}{\partial s}, \Theta_s^{(k-1)}\right)\right\} + \text{ exact form on } M.$$

Proof. Use (4.2.35) for the link of the connections θ_0, θ_s $(s \in I)$ given by

$$\tau = \theta_{\sigma s} + s\lambda(p,\sigma s)d\sigma \qquad (p \in P;\ 0 \leq \sigma \leq 1) .$$

The result is

$$T(\theta_s)f = T(\theta_0)f + k\int_0^s f\left(\frac{\partial\theta_u}{\partial u} - [\theta_u,\lambda(p,u)] - d_P\lambda(p,u)\right. ,$$

$$\left.\Theta_u^{(k-1)}\right) du + \text{exact form on } P =$$

$$= T(\theta_0)f + k\int_0^s f\left(\frac{\partial\theta_u}{\partial u}\ ,\ \Theta_u^{(k-1)}\right) du + \text{exact form on } P,$$

and, by derivating with respect to s we get (4.2.40). Formula (4.2.41) follows similarly from (4.2.36). Q.e.d.

The derivation formulas (4.2.40) and (4.2.41) generalize derivation formulas established by Chern-Simons [18] and J.L. Heitsch [39] respectively. The original formulas were for $\lambda=0$, but taking $\lambda\neq 0$ is useful, for instance, in the important Theorem 4.2.16 later on.

Another interesting result is

Proposition 4.2.14. *Let* θ_0, θ_1 *be two connections on* P, *and* f *an invariant polynomial such that* $f \in \ker\Delta(\theta_0) \cap \ker\Delta(\theta_1)$. *Then we have the following relation between Chern-Simons and secondary characteristic classes*

$$\pi^*[\Delta(\theta_0,\theta_1)f] = [T(\theta_1)f] - [T(\theta_0)f] ,$$

where $\pi: P \to M$ *is the projection of the bundle* P.

Proof. Formula

(4.2.43) $$\varphi = \theta_0 + s(\theta_1 - \theta_0) = \theta_0 + s\alpha \qquad (\alpha = \theta_1 - \theta_0)$$

provides a link between θ_0 and θ_1, and, for this link, (4.2.35) becomes

(4.2.44) $$T(\theta_1)f - T(\theta_0)f = \pi^*\Delta(\theta_0,\theta_1)f + \text{exact form}.$$

Q.e.d.

Now, following D. Lehmann [49] we define

Definition 4.2.15. *Let* $f \in I^k(G)$, *and* θ_0, θ_1 *be G-connections on* P. *Then, these two connections are called* **f-homotopic** *if there is a link* φ *between* θ_0 *and* θ_1 *such that* $f \in \ker \Delta(\varphi)$.

Accordingly, we can prove

Theorem 4.2.16. *The Chern-Simons and secondary characteristic classes defined by a polynomial* $f \in I^k(G)$ *for a G-principal bundle* $\pi: P \to M$ *remain unchanged if the connections used in their definition are replaced by f-homotopic connections.*

Proof. Indeed, if θ_0 and θ_1 are f-homotopic by φ, then (4.2.38) and $f \in \ker \Delta(\varphi)$ imply the vanishing of the integral term of the first equality (4.2.35), and (4.2.35) yields $[T(\theta_0)f] = [T(\theta_1)f]$. Then, if θ_0' and θ_1' are also f-homotopic connections by a link φ', the integral term in the first equality (4.2.36) vanishes again by (4.2.38), and (4.2.36) yields $[\Delta(\theta_0,\theta_0')f] = [\Delta(\theta_1,\theta_1')f]$. Q.e.d.

Corollary 4.2.17. *If* θ_0 *and* θ_1 *are f-homotopic connections, the cohomology class* $[\Delta(\theta_0,\theta_1)f] = 0$.

Proof. If the f-homotopy is realized by a link φ, then $f \in \ker \Delta(\varphi)$ implies $f \in \ker \Delta(\theta_0) \cap \ker \Delta(\theta_1)$, and it follows from Theorem 4.2.16 that $[\Delta(\theta_0,\theta_1)f] = [\Delta(\theta_1,\theta_1)f] = 0$. Q.e.d.

Now, we shall give one more result about Chern-Simons classes useful in later applications. In order to formulate it, we remember the universal Chern-Weil homomorphism $w(G)\colon I(G) \to H^*(B_G,\mathbb{R})$ of (4.2.22) (remember also that grade $f = 2k$ if $f \in I^k(G)$). Then, we may define the sets

$$(4.2.45) \qquad I_0^k(G) = \{f \in I^k(G)/w(G)(f) \in r(H^{2k}(B_G,\mathbb{Z}))\},$$

where $r\colon H^{2k}(B_G,\mathbb{Z}) \to H^{2k}(B_G,\mathbb{R})$ is the *real image map* induced by $\mathbb{Z} \subset \mathbb{R}$.

Proposition 4.2.18. [18]. *Let* $f \in I_0^k(G)$, *and let* θ *be a connection of the* G-*principal bundle* $\pi\colon P \to M$ *such that* $f \in \ker \Delta(\theta)$. *Then there exists a cohomology class* $u \in H^{2k-1}(M,\mathbb{R}/\mathbb{Z})$ *such that*

$$(4.2.46) \qquad [T(\theta)f] \ (\text{mod.}\ \mathbb{Z}) = \pi^*u ,$$

where mod. $\mathbb{Z}$ *denotes the image of a real cohomology class under the natural projection* $\mathbb{R} \to \mathbb{R}/\mathbb{Z}$.

Proof. We start with a few remarks. The exact sequence

$$0 \to \mathbb{Z} \overset{\subset}{\to} \mathbb{R} \overset{pr}{\to} \mathbb{R}/\mathbb{Z} \to 0$$

has the associated exact cohomology sequence

$$\ldots \to H^j(X,\mathbb{Z}) \overset{r}{\to} H^j(X,\mathbb{R}) \overset{\text{mod}\ \mathbb{Z}}{\longrightarrow} H^j(X,\mathbb{R}/\mathbb{Z}) \to H^{j+1}(X,\mathbb{Z}) \to \ldots$$

for any space X, which shows that $\alpha \in r(H^j(X,\mathbb{Z}))$ iff $\alpha(\text{mod.}\mathbb{Z}) = 0$. A differential form λ defines a real cochain by integration along chains c. Similarly $\int_c \lambda$ (mod $\mathbb{Z}$) define an $\mathbb{R}/\mathbb{Z}$-cochain, which we shall denote by $\tilde{\lambda}$. Particularly, if $d\lambda = 0$, $[\tilde{\lambda}] = [\lambda]$(mod $\mathbb{Z}$).

Now, let $(\pi_G\colon E_G^N \to B_G^N, \theta_G^N)$ be a N-universal bundle with connection, in the sense of Theorem 4.1.3, and with a "very large" N comparative to k. Then, the cohomology of B_G^N is equal to that of B_G, in dimension k, and we have

$$(4.2.47) \qquad w(G,E_G)(f) = [\Delta(\theta_G^N)f] = r(\alpha) \ , \quad \alpha \in H^{2k}(B_G^N,\mathbb{Z}) \ .$$

Hence, $[\Delta(\theta_G^N)f]$ (mod. $\mathbb{Z}$) = 0 and

(4.2.48) $$\widetilde{\Delta(\theta_G^N)f} = dv ,$$

where v is a (2k-1)-dimensional $\mathbb{R}/\mathbb{Z}$-cochain of B_G^N, and d is the $\mathbb{R}/\mathbb{Z}$-coboundary.

From (4.2.30), and (4.2.48) we have

$$d\left(\widetilde{T(\theta_G^N)f}\right) = \pi_G^*\left(\widetilde{\Delta(\theta_G^N)f}\right) = d(\pi_G^* v) .$$

But, another known result of topology says that E_G^N is acyclic in dimension 2k-1 [14] (since N is large). Therefore we get

(4.2.49) $$\widetilde{T(\theta_G^N)f} = \pi_G^* v + (\mathbb{R}/\mathbb{Z}\text{-coboundary in } E_G^N).$$

Furthermore, if $\varphi: M \to B_G$ is a classifying mapping for the connection θ, then by applying φ^* to (4.2.49) we get

(4.2.50) $$\widetilde{T(\theta)f} = \pi^*(\varphi^* v) + (\mathbb{R}/\mathbb{Z}\text{-coboundary in } P) .$$

Since we also have the hypothesis $\Delta(\theta)f = 0$, we get from (4.2.30) that $d\left(\widetilde{T(\theta)f}\right) = 0$ and then (4.2.50) yields $\pi^*\left(d(\varphi^* v)\right) = 0$. Since π^* is injective at the cochain level this shows that $\varphi^* v$ is an $\mathbb{R}/\mathbb{Z}$-cocycle, and it defines a cohomology class $u = [\varphi^* v] \in H^{2k-1}(M, \mathbb{R}/\mathbb{Z})$. The requested result is then an immediate consequence of (4.2.50). Q.e.d.

Remark 4.2.19. *The concepts of Proposition 4.2.18 have been generalized by J. Simons and J. Cheeger into a rather ample theory of so-called* ***differential characters*** *which complements the theory of characteristic classes* [73], [17].

Now, we shall introduce a general definition of secondary characteristic classes. We shall follow D. Lehmann's paper [49]. Lehmann's theory is based on the consideration of proper *homogeneous ideals* J of the algebra $I(G)$, which means that J is generated by homogeneous polynomials, and $J \subset J^+(G) = \bigoplus_{k \geq 1} I^k(G)$.

Definition 4.2.20. *Let* J_0, J_1 *be two homogeneous ideals in* $I(G)$. *Then, the algebra*

$$W(J_0, J_1) = [I(G)/_{J_0}] \otimes_{\mathbb{R}} [I(G)/_{J_1}] \otimes_{\mathbb{R}} \Lambda(I^+(G)) \quad , \tag{4.2.51}$$

where the first two factors are quotient algebras, and the last factor is the exterior algebra generated by the vector space $I^+(G)$, *will be called a* ***secondary universal algebra of*** G.

Furthermore, we shall attribute *grades* to the elements of $W(J_0, J_1)$ by means of the rules

$$\text{grade } [f]_0 = \text{grade } [f]_1 = 2k, \ \text{grade } \hat{f} = 2k-1 \ , \tag{4.2.52}$$

where $f \in I^k(G)$, and $[f_0]$, $[f_1]$, $\hat{f}$ denote the elements defined by f in each factor of (4.2.51), respectively.

We shall also define a differentiation by

$$d[f\,]_0 = d[f]_1 = 0 \ , \ d\hat{f} = [f]_1 - [f]_0 \tag{4.2.53}$$

and by the natural algebraic extensions of these operations. d raises the grades by 1, and $d^2 = 0$.

Thereby, $W(J_0, J_1)$ becomes a *differential graded algebra* which has a corresponding *cohomology algebra* $H^*\big(W(J_0, J_1)\big)$ defined in the usual way (like for the usual differential forms) [34].

Now, let $\pi: P \to M$ be a G-principal bundle, and let θ_a be connections of P such that $J_a \subset \ker \Delta(\theta_a)$ $(a = 0,1)$. Such a connection θ_a is called a J_a*-connection*. Then, we may define a homomorphism

(4.2.54) $\rho(\theta_0,\theta_1) : W(J_0,J_1) \to \Lambda(M)$

by the natural algebraic extension of the following formulas

(4.2.55) $\rho(\theta_0,\theta_1)[f]_0 = \Delta(\theta_0)f \ , \ \rho(\theta_0,\theta_1)[f]_1 = \Delta(\theta_1)f \ ,$

$$\rho(\theta_0,\theta_1)\hat{f} = \Delta(\theta_0,\theta_1)f \ .$$

Definition 4.2.21. *The homomorphism (4.2.54) is called the* ***secondary Chern-Weil homomorphism.*** *(A universal secondary Chern-Weil homomorphism can also be defined [49, Appendix by B. Callenaere and D. Lehmann]).*

Clearly, the homomorphism $\rho(\theta_0,\theta_1)$ is grade preserving, and because of (4.2.16) and (4.2.18) ρ commutes with the differentials d of the two algebras involved in (4.2.54). Hence, there is an induced *cohomological secondary Chern-Weil homomorphism*

(4.2.56) $\rho^*(\theta_0,\theta_1) : H^*(W(J_0,J_1)) \to H^*(M,\mathbb{R})$.

The restriction of ρ^* to $I(G)/J_a$ $(a = 0,1)$ is the natural factorization of the Chern-Weil homomorphism $w(G,P)$, which allows us to give

Definition 4.2.22. *The cohomology classes sitting in* $\operatorname{im} \rho^*(\theta_0,\theta_1) \smallsetminus \operatorname{im} w(G,P)$ *are called* ***secondary characteristic classes*** *of* (P,θ_0,θ_1).

The following result is now immediate

Proposition 4.2.23. *Any simple secondary characteristic class of* (P,θ_0,θ_1) *is also a secondary class in the sense of Definition 4.2.22.*

Proof. If the simple class is defined by $f \in I^k(G)$ then the same class in the new definition is given by $\hat{f}$. Q.e.d.

Another interesting computation holds for the case of a polynomial $f = uv$, where $u \in I^h(G)$, $v \in I^\ell(G)$, $h+\ell = k$. Then, the element

$$(4.2.57) \qquad \xi = \widehat{uv} - [u]_0 \otimes \hat{v} - [v]_1 \otimes \hat{u}$$

satisfies $d\xi = 0$, and it is a cocycle of $W(J_0,J_1)$. (Just a straightforward application of (4.2.53)). However, ξ give no secondary classes since we have [49]

Proposition 4.2.24. *For every* ξ *given by (4.2.57),* $\rho^*(\theta_0,\theta_1)[\xi] = 0$.

Proof. We have $\rho(\theta_0,\theta_1)\widehat{uv} = \Delta(\theta_0,\theta_1)(uv)$, and it is convenient to write this form using the equivalent form (4.2.11) of (4.2.26), while observing the notation (4.2.24), (4.2.25). We get

$$(4.2.58) \qquad \rho(\theta_0,\theta_1)\widehat{uv} = -\int_{\Delta^1} (uv)\left(\tilde{\Theta}^{(k)}\right) = -\int_{\Delta^1} u\left(\tilde{\Theta}^{(h)}\right) \wedge v\left(\tilde{\Theta}^{(\ell)}\right) =$$

$$= \int_0^1 \left\{ku\left(\alpha,\Theta_t^{(h-1)}\right) \wedge v\left(\Theta_t^{(\ell)}\right) + \ell u\left(\Theta_t^{(h)}\right) \wedge v\left(\alpha,\Theta_t^{(\ell-1)}\right)\right\} dt \; .$$

Now, the derivation formula (4.2.21) yields

$$(4.2.59) \qquad \begin{aligned} \frac{\partial\left(\Delta(\theta_0,\tilde{\theta})u\right)}{\partial t} &= ku\left(\alpha,\Theta_t^{(h-1)}\right) + \text{exact form on } M, \\ \frac{\partial\left(\Delta(\theta_1,\tilde{\theta})v\right)}{\partial t} &= \ell v\left(\alpha,\Theta_t^{(\ell-1)}\right) + \text{exact form on } M \quad , \end{aligned}$$

and since $u\left(\Theta_t^{(h)}\right)$, $v\left(\Theta_t^{(\ell)}\right)$ are closed forms, (4.2.58) modulo (4.2.59) can be interpreted as

$$(4.2.60) \qquad \rho(\theta_0,\theta_1)\widehat{uv} = -\int_I d\Lambda + \text{exact form on } M \; ,$$

where Λ is the form defined on $M \times I$ by

$$\Lambda = u\left(\Theta_t^{(h)}\right) \wedge [\Delta(\theta_1,\theta_t)v] + v\left(\Theta_t^{(\ell)}\right) \wedge [\Delta(\theta_0,\theta_t)u] \; .$$

Finally, let us apply Stokes' formula (4.1.29) to the integral (4.2.60).

We shall obtain

$$(4.2.61)\qquad \rho(\theta_0,\theta_1)\widehat{uv} = u\left(\Theta_0^{(h)}\right) \wedge [\Delta(\theta_0,\theta_1)v] +$$

$$+ v\left(\Theta_1^{(h)}\right) \wedge [\Delta(\theta_0,\theta_1)u] + \text{exact form on } M =$$

$$= \{\rho(\theta_0,\theta_1)[u]_0\} \wedge \{\rho(\theta_0,\theta_1)\hat{v}\} +$$

$$+ \{\rho(\theta_0,\theta_1)[v]_1\} \wedge \{\rho(\theta_0,\theta_1)\hat{u}\} + \text{exact form on } M .$$

Q.e.d.

Remark 4.2.25. *This shouldn't give us the impression that there are no nonsimple secondary classes. Indeed, there are plenty of such classes, for instance in foliation theory* [*49*].

The following theorem [49] is important for the computation of characteristic classes

Theorem 4.2.26. *Let* A *be a system of homogeneous generators of* $I^+(G)$, *let* $\wedge(A)$ *be the exterior algebra spanned by these generators, and let us denote*

$$(4.2.62)\qquad W_A(J_0,J_1) = [I(G)/_{J_0}] \otimes_{\mathbb{R}} [I(G)/_{J_1}] \otimes_{\mathbb{R}} \wedge(A) .$$

Then, if $\rho_A(\theta_0,\theta_1)$ *is the restriction of* $\rho(\theta_0,\theta_1)$ *to* $W_A(J_0,J_1)$ *one has* $\operatorname{im} \rho_A^*(\theta_0,\theta_1) = \operatorname{im} \rho^*(\theta_0,\theta_1)$.

Proof. That A generates $I^+(G)$ means every $f \in I^+(G)$ is a linear combination of products $a_1 a_2,\ldots,a_r$ in I(G) where $a_h \in A$ $(h=1,2,\ldots,r)$. Now let us see what happens if $x \in W(J_0,J_1)$ is a cocycle that contains a product $\hat{f} = \widehat{uv}$ $(f \in I^+(G))$. Then

$$(4.2.63)\qquad x = \hat{f} \wedge x' + x'' ,$$

where x',x'' do not contain $\hat{f}$ any more, and

$$(4.2.64) \qquad dx = ([f]_1 - [f]_0) \otimes x' + dx'' - \hat{f} \wedge dx' = 0.$$

Clearly, (4.2.64) holds iff

$$(4.2.65) \qquad ([f]_1 - [f_0]) \otimes x' + dx'' = 0 \quad , \quad dx' = 0 \ .$$

Now, with a look at (4.2.57), let us define

$$(4.2.66) \qquad y = ([u]_0 \otimes \hat{v} + [v]_1 \otimes \hat{u}) \wedge x' + x'' \quad ,$$

and notice that (4.2.65) implies $dy = 0$. On the other hand, we have $x\text{-}y = \xi \wedge x'$, with ξ defined by (4.2.57), and since $\rho(\theta_0\theta_1)\xi$ is an exact form by Proposition 4.2.24, while $\rho(\theta_0,\theta_1)x'$ is a closed form by (4.2.65), we get

$$(4.2.67) \qquad \rho(\theta_0,\theta_1)y = \rho(\theta_0,\theta_1)x + \text{exact form} \ .$$

This implies $\rho^*(\theta_0,\theta_1)y = \rho^*(\theta_0,\theta_1)x$. But y contains only the "simple factors" $\hat{u}$, $\hat{v}$. Hence, generally, we may "reduce" the products $a_1 a_2,\ldots,a_r$ mentioned at the beginning of this proof to simple factors, and get an element of $W_A(J_0,J_1)$ whose image by ρ^* is the same as the image of the original element of $W(J_0,J_1)$. Q.e.d.

In other words, the computation of all the secondary characteristic classes can be done with the subalgebra $W_A(J_0,J_1)$ alone.

Like for the simple secondary classes, we must also discuss the dependence of the general secondary classes on the choice of the connections. First we give [49]

Definition 4.2.27. *Let* J *be a homogeneous ideal in* $I(G)$. *Then two connections* θ_0,θ_1 *on* P *are called* ***differentially*** J-***homotopic*** *if there is a link* φ *between* θ_0 *and* θ_1 *such that* $J \subset \ker \Delta(\varphi)$. *Furthermore, two connections* θ_0,θ_1 *are called* J-***homotopic*** *if there is a finite chain* *of connections* $\theta_0, \theta^{(1)}, \ldots, \theta^{(s)}, \theta_1$ *such that any two consecutive connections of this chain are differentiably J-homotopic.*

The relation of J-homotopy is an equivalence relation, and it divides the set of the connections of P into disjoint *J-homotopy classes of connections.*

Now, we can easily establish

Theorem 4.2.28. *The cohomological secondary Chern-Weil homomorphism* $\rho^*(\theta_0, \theta_1)$ *remains unchanged if the connections* θ_0, θ_1 *are replaced by* θ_0', θ_1' *where* θ_a *is* J_a*-homotopic to* θ_a' $(a=0,1)$.

Proof. Let φ_0, φ_1 be the links that define the J_a-homotopies of the statement (it suffices, of course, to assume that these homotopies are differentiable in the sense of Definition 4.2.27). Take $x \in W(J_0, J_1)$, and apply Stokes' formula (4.1.29) to $\alpha = \rho(\varphi_0, \varphi_1)x$ on $M \times I$. We get

$$d \int_I \rho(\varphi_0, \varphi_1)x - \int_I d\rho(\varphi_0, \varphi_1)x = \tag{4.2.68}$$

$$= (-1)^{\text{grade } x - 1} \{\rho(\theta_0', \theta_1')x - \rho(\theta_0, \theta_1)x\} .$$

This implies $\rho^*(\theta_0, \theta_1) = \rho^*(\theta_0', \theta_1')$. Q.e.d.

In other words, a couple of ideals (J_0, J_1), and a couple C_a of J_a-homotopy classes of connections $(a=1,0)$ yields secondary characteristic classes which do not depend on the choice of the connections within the classes C_a.

Corollary 4.2.29. *With the notation above, if* $C_0 \cap C_1 \neq \phi$, *all the secondary characteristic classes vanish.*

Proof. Because of Theorem 4.2.28, we may replace the connections θ_0, θ_1 by $\theta_0' = \theta_1' \in C_0 \cap C_1$. Then the secondary classes vanish because of $\Delta(\theta_0', \theta_1')f = 0$. Q.e.d.

Hence, the existence of nonzero secondary classes implies $C_0 \cap C_1 = \phi$, and *the secondary characteristic classes have the geometric meaning of being obstructions to the existence of a common connection in the classes* C_0, C_1.

The reader may find various applications of the theory of the secondary characteristic classes in works such as [49], [43], etc.

Let us end this Section by a short exposition of a subject intimately related with characteristic classes namely the *residues theory* as developed by D. Lehmann [50].

Let $\pi: P \to M$ be a G-principal bundle naturally equiped with a *connection with singularities* θ, i.e., a connection defined on $P/_{M \setminus S}$, where S is a closed subset of M called the *singular set.* Let us also assume that for a given integer k, and for some differentiable trangulation K of M one has $S \cap (\text{skeleton}^{2k-1} K) = \phi$. Take $f \in I^k(G)$, and take a 2k-simplex γ of K. Since γ is a contractible set, there is a section $\sigma: \gamma \to P$, and a connection θ_0 on $P/_\gamma$ given by the local connection form $\sigma^*\theta_0 = 0$. We call such a θ_0 a *trivial connection.*

Definition 4.2.30. *The integral*

$$\text{(4.2.69)} \qquad \text{Res}_\gamma(\theta, f) = \int_{\partial\gamma} \Delta(\theta_0, \theta) f$$

is called the ***f-residue of*** θ ***along*** γ.

Notice that the integrand of the right-hand side of (4.2.69) may not be defined on γ, because of the existence of singularities.

Proposition 4.2.31. (The residues formula [50]). *If* $\bar{\theta}$ *is an arbitrary G-connection without singularities on* P, *then*

$$\text{(4.2.70)} \qquad \text{Res}_\gamma(\theta, f) = \int_\gamma \Delta(\bar{\theta}) f - \int_{\partial\gamma} \Delta(\theta, \bar{\theta}) f \, .$$

Proof. It is known that connections $\bar{\theta}$ always exist. From Bott's formula (4.2.13), we get

$$d\Delta(\theta_0,\theta,\bar{\theta})f = \Delta(\theta,\bar{\theta})f - \Delta(\theta_0,\bar{\theta})f + \Delta(\theta_0,\theta)f \; .$$

If we integrate this relation along $\partial\gamma$, and apply the usual Stokes' theorem we get

$$\int_{\partial\gamma} \Delta(\theta_0,\theta)f = \int_{\partial\gamma} \Delta(\theta_0,\bar{\theta})f - \int_{\partial\gamma} \Delta(\theta,\bar{\theta})f \; . \tag{4.2.71}$$

Then, in view of (4.2.18)

$$\int_{\partial\gamma} \Delta(\theta_0,\bar{\theta})f = \int_{\gamma} d\Delta(\theta_0,\bar{\theta})f = \int_{\gamma} \Delta(\bar{\theta})f - \int_{\gamma} \Delta(\theta_0)f \; .$$

But the trivial connection θ_0 has a vanishing curvature. Hence $\Delta(\theta_0)f = 0$, and (4.2.71) is precisely the requested formula. Q.e.d.

Corollary 4.2.32. *The residue (4.2.69) does not depend on the choice of the local trivial connection* θ_0 .

Proof. Obvious, from (4.2.70). Q.e.d.

Now, let $\ell = \Sigma_q a_q \gamma_q$ be a real 2k-dimensional chain of K. Then it is natural to define the *sum of* f-*residues of* θ *along* ℓ by

$$\mathrm{Res}_{\ell}(\theta,f) = \sum_q a_q \, \mathrm{Res}_{\gamma_q}(\theta,f) \; . \tag{4.2.72}$$

As a consequence of the residues formula we get

Theorem 4.2.33. (The residues theorem). *For every* 2k-*dimensional cycle* c *of* K, *one has*

$$\mathrm{Res}_c(\theta,f) = \int_c \Delta(\bar{\theta})f = \langle [\Delta(\bar{\theta})f],c \rangle \; , \tag{4.2.73}$$

where $\bar{\theta}$ *is an arbitrary connection without singularities on* P.

Proof. If $\mathrm{Res}_c(\theta,f)$ is evaluated by means of (4.2.70), we get

$$\mathrm{Res}_c(\theta,f) = \int_c \Delta(\bar{\theta})f - \int_{\partial c} \Delta(\theta,\bar{\theta})f .$$

But the last integral vanishes since $\partial c = 0$. Q.e.d.

The last term of (4.2.73) is the value of the characteristic class defined by f on the chain c. This shows the relation between residues and characteristic classes. The residues theorem has such applications as the Gauss-Bonnet theorem and Hopf's theorem on singularities of vector fields for surfaces, the Riemann-Hurwitz formula, etc. [50].

As a matter of fact, the residues generalize the construction of characteristic classes by assigning cohomology classes to connections with singularities as well. Indeed, if the chain ℓ of (4.2.72) is seen as a varying argument, then formula (4.2.72) yields a 2k-dimensional cochain $\mathrm{Res}(\theta,f)$, and it is a straightforward consequence of (4.2.70), and of the classical Stokes' theorem that $\mathrm{Res}(\theta,f)$ is in fact a cocycle. Hence, we get a cohomology class $[\mathrm{Res}(\theta,f)] \in H^{2k}(K,\mathbb{R}) \approx H^{2k}(M,\mathbb{R})$. In case the connection θ has no singularities, we may use Stokes' theorem and formula (4.2.18) in (4.2.69), and we see that $[\mathrm{Res}(\theta,f)] = [\Delta(\theta)f]$, which is a primary characteristic class.

The residues are also related to the secondary characteristic classes. Let us assume that $M \setminus S$ retracts to a subcomplex $K' \subset K$. On the other hand, let us assume that $f \in \ker \Delta(\theta)$ on $M \setminus S$. Then (4.2.69) yields that for $\gamma \in K'$ one has

$$\mathrm{Res}_\gamma(\theta,f) = \int_{\partial\gamma} \Delta(\theta_0,\theta)f = \int_\gamma d\Delta(\theta_0,\theta)f = 0 ,$$

and $\mathrm{Res}(\theta,f)$ is a relative cochain of K with respect to K'. Now, let us denote

(4.2.74) $\qquad C(f) = \{\bar{\theta}/\bar{\theta}$ is a connection on P and $f \in \ker \Delta(\bar{\theta})\}$,

and use a fixed $\bar{\theta} \in C(f)$ in (4.2.70). Then we get

$$(4.2.75) \qquad \mathrm{Res}_\gamma(\theta,f) = -\int_{\partial\gamma} \Delta(\theta,\bar\theta)f = -\langle [\Delta(\theta,\bar\theta)f],\partial\gamma\rangle$$

which is meaningful since $\Delta(\theta,\bar\theta)f$ is a closed form on $M \setminus S$, and $\partial\gamma \in K'$ because of $S \cap \mathrm{skeleton}^{2k-1}K = \phi$. Formula (4.2.75) proves

Proposition 4.2.34. *Under the hypotheses described above,* $-[\mathrm{Res}(\theta,f)]$ *is the image of the simple secondary characteristic class* $[\Delta(\theta,\bar\theta)f]$ *of* $M \setminus S$, *under the connecting homomorphism of the exact cohomology sequence of the pair* (K,K') [75].

Of course, in this Proposition $\bar\theta$ can vary arbitrarily in $C(f)$. We also see that if θ and $\bar\theta$ are f-homotopic over $M \setminus S$, then the residues of θ are necessarily zero. Hence $\mathrm{Res}(\theta,f)$ is an obstruction to the existence of a global f-homotopic extension of θ.

4.3. THE MASLOV CLASS AND INDEX. We shall want to study the characteristic classes of symplectic bundles and Lagrangian subbundles. One such class has been discovered in the study of asymptotic solutions of linear partial differential equations, and it is called *the Maslov class* see Section 1.1). The original viewpoint on this class was different and Maslov's class was studied in many ways of a great geometrical and topological interest [57], [2], [29], [51], [74], [41], [61], [23], etc. For this reason, we make an intermission, and we introduce the present survey section, where we indicate *without proofs* some of the most important approaches to the theory of the Maslov class. The interested reader is referred to the previously quoted works as well as to the ample survey of [37].

Let $(\pi\colon E \to M^m,\omega)$ be a symplectic vector bundle of rank 2n, and let L_0, L_1 be Lagrangian subbundles of E (see Section 3.1 for definitions and notation). If we are interested in studying the transversality of L_0 and L_1, we must consider various relations that can be established between two such subbundles.

The most straightforward is the relation between local bases determined by L_0 and L_1. Since, by Theorem 3.1.2, the structure group of E can be reduced to $U(n)$, it is natural to consider such a reduction defined by a positive compatible complex structure J of E, with the associated hermitian metric g. Then, if $U \subset M$ is a trivializing neighbourhood of the bundles considered, we have fields of L_a-related unitary frames $(a=0,1)$ (see Definition 2.3.8) over U, and the complex form of these frames is (formula (2.3.13))

$$\varepsilon_i^{(a)} = \frac{1}{\sqrt{2}}\left(e_i^{(a)} - \sqrt{-1}\, Je_i^{(a)}\right) , \tag{4.3.1}$$

where $\left(e_i^{(a)}\right)$ is a g-orthonormal basis of L_a $(a=0,1)$. This implies a relationship

$$\varepsilon^{(1)} = \varepsilon^{(0)} A_U \tag{4.3.2}$$

written in matrix form, where $A_U\colon U \to U(n)$ is a differentiable mapping. A change of the L_a-related bases in (4.3.2) multiplies A_U at the left and at the right by orthogonal matrices. Hence, over a trivializing covering $\{U\}$ of M, $\{A_U\}$ glue up to a global set mapping

$$A(L_0,L_1)\colon M \to U(n) \| O(n) , \tag{4.3.3}$$

where $\|$ denotes equivalence of unitary matrices under both left and right multiplication by orthogonal matrices. We call $A(L_0,L_1)$ the *connecting mapping* of L_0 and L_1, and it depends on the choice of J.

Now, the connecting mapping has an obvious invariant namely, $\det^2 A(L_0,L_1) = \det^2 A_U$, which is well defined since the determinant of an orthogonal matrix is ± 1 (det means determinant). More exactly, since the determinant of a unitary matrix has modulus 1, we have a well defined differentiable mapping

$$(4.3.4) \qquad \varphi(L_0, L_1) = \det^2 \circ A(L_0, L_1) : M \to S^1 \quad ,$$

where S^1 is the unit circle, and we may give [3]

Definition 4.3.1. *Let* $dz/2\pi\sqrt{-1}\, z$ $\left(z \in S^1 = \{z \in \mathbb{C} / |z| = 1\}\right)$ *be the (usual) basis of 1-forms of* $S^1 = \mathbb{R}/\mathbb{Z}$. *Then, the cohomology class*

$$(4.3.5) \qquad m(L_0, L_1) = \varphi^*(L_0, L_1) \left[\frac{dz}{2\pi\sqrt{-1}z}\right] \in H^1(M, \mathbb{R})$$

is called the ***Maslov class*** *of* (L_0, L_1), *and the number*

$$(4.3.6) \qquad m^{\gamma}_{L_0 L_1} = \int_{\gamma} m(L_0, L_1) = \deg\left(\varphi(L_0, L_1) \circ \gamma\right) \quad ,$$

where $\gamma : S^1 \to M$ *is a closed curve, and* deg *denotes the degree of the mapping, is called the* ***Maslov index*** *of* γ.

Proposition 4.3.2. *The Maslov class is an integral cohomology class, and the Maslov index is an integer. They do not depend on the choice of the compatible structure* J, *and vanish if* L_0 *and* L_1 *are transversal Lagrangian subbundles.*

Proof. Here, by an integral class, we understand the real image of an integral cohomology class, and, since $[dz/2\pi\sqrt{-1}\, z]$ is an integral class, the first assertion is trivial. The independence of $m(L_0, L_1)$ on J follows from the fact that two structures J_0, J_1 are always connected by a homotopy J_t $(0 \leq t \leq 1)$ (Theorem 3.1.2), hence the corresponding mapping φ of (4.3.4) will be homotopic. (It is for this reason that we use cohomology classes, and not just forms, in (4.3.5).) Finally, if L_0 and L_1 are transversal subbundles, we choose J such that $L_1 = JL_0$ which yields $A_U = \sqrt{-1}\, I_n$ (I_n = the unit n-matrix), and (4.3.5) gives $m(L_0, L_1) = 0$. Q.e.d.

Therefore, in this way we have obtained a basic transversality obstruction for Lagrangian subbundles.

Examples 4.3.3. a) Take the symplectic manifold $(\mathbb{R}^2=\{(x,y)\},\omega=dx\wedge dy)$, and let γ be a simple closed curve in $\mathbb{R}^2$. Then γ is a Lagrangian submanifold, and (Example 1.1.5 a)) $E = T\mathbb{R}^2/\gamma$ is a symplectic vector bundle, where $L_1 = T\gamma$ is a Lagrangian subbundle of the former. Another Lagrangian subbundle is L_0 defined by the lines y=const. along γ. Then, if θ denotes the angle between L_0 and L_1, the bases of L_1 and L_0 are related by

$$e^{(1)} = e^{(0)}\cos\theta + Je^{(0)}\sin\theta ,$$

where J is the complex structure given by $z = x + \sqrt{-1}\,y$ in $\mathbb{R}^2$. We deduce from (4.3.1)

$$\varepsilon^{(1)} = e^{i\theta}\varepsilon^{(0)}$$

and, from (4.3.6), it follows that $m^{\gamma}_{L_0L_1}$ is twice the rotation index of the tangent of γ. Since γ is a simple curve, $m^{\gamma}_{L_0L_1} = \pm 2$ according to the orientation of γ. If the closed curve γ is not simple we may look at it as an immersion in $\mathbb{R}^2$, and the pullback by this immersion yields a similar configuration with $m^{\gamma}_{L_0L_1}$ = twice the rotation index of the tangent of $\gamma = \frac{1}{\pi}\int_{\gamma}\kappa\, ds$, where κ is the curvature, and s is the arc length along γ. Notice that this example yields cases of nonvanishing Maslov classes. On the contrary, we may have a curve γ of vanishing rotation index of the tangents but which is not transversal to y=const. (a figure eight curve, for instance). Hence the vanishing of the Maslov class is not enough for transversality.

b) Consider the unit sphere $S^3 \subset \mathbb{R}^4$ with the coordinates x^i $(i = 1,2,3,4)$. On S^3, one has the tangent vector fields

$$(4.3.7) \qquad \xi_1 = (-x^2, x^1, -x^4, x^3),\ \xi_2 = (-x^3, x^4, x^1, -x^2)$$
$$\xi_3 = (-x^4, -x^3, x^2, x^1) \quad ,$$

which define a *parallelization* of the sphere. Let (η_1, η_2, η_3) be the dual cotangent basis, where

$$(4.3.8) \qquad \eta_1 = -x^2 dx^1 + x^1 dx^2 - x^4 dx^3 + x^3 dx^4 \quad ,$$

etc. The forms η_a $(a = 1,2,3)$ are easily seen to be contact forms on S^3, and we look at the contact manifold (S^3, η_1) (Definition 3.2.10). Its contact distribution is $C_1 = \text{span}\ \{\xi_2, \xi_3\}$, and $(C_1, d\eta_1|_{C_1})$ is a symplectic vector bundle of rank 2. A curve γ of S^3 which is always tangent to C_1 will be called a ξ_1*-Legendre curve*, and it is a Legendrian submanifold of S^3 (Definition 3.3.13). Hence $L_1 = T\gamma = \text{span}\ \{\dot{\gamma}\}$ is a Lagrangian subbundle of $C_1|_\gamma$. On the other hand, $\xi_2|_\gamma$ is another Lagrangian subbundle L_0. It is easy to see that

$$(4.3.9) \qquad J\xi_2 = \xi_3 \ ,\ J\xi_3 = -\xi_2$$

defines a compatible complex structure in C_1. Furthermore, if θ is the angle of $\dot{\gamma}$ with ξ_2, we have

$$(4.3.10) \qquad \dot{\gamma} = \xi_2 \cos\theta + \xi_3 \sin\theta,\ J\dot{\gamma} = -\xi_2 \sin\theta + \xi_3 \cos\theta \ ,$$

and if we take $e^{(0)} = \xi_2$, $e^{(1)} = \dot{\gamma}$ in (4.3.1), we obtain

$$(4.3.11) \qquad \varepsilon^{(1)} = \exp[\sqrt{-1}\ \theta]\varepsilon^{(0)} \ .$$

Hence, like in Example a), if γ is closed, $m^{\gamma}_{L_0 L_1}$ is twice the rotation index of the angle θ.

As we said already, the Maslov class and index were studied in many different ways. Originally, there was a *Maslov index* that appeared in studying a Lagrangian submanifold M^n of the standard symplectic space

$\mathbb{R}^{2n}$ for the purpose of constructing asymptotic solutions of equations (1.1.8) [57], [58], [2]. Namely, let (x^i, y^i) $(i = 1,\ldots,n)$ be the coordinates of $\mathbb{R}^{2n}$ ($\omega = \Sigma_{i=1}^{n} dx^i \wedge dy^i$ is the standard symplectic form), and f the projection of M onto the plane $y^i = 0$. Let $\Sigma \subset M$ be the set of singular points of f. In the "generic case", which can always be achieved by a "small deformation", Σ defines a (n-1)-dimensional two-sided cycle of M for which a positive and a negative side can be conveniently chosen. Now, if γ is a path in M with nonsingular endpoints, and transversal to Σ, one defines index $\gamma = \nu_+ - \nu_-$, where ν_+ is the number of points where γ passes from the negative to the positive side of Σ, and ν_- is the number of points where γ passes from the positive to the negative side. V.I. Arnold proved that if γ is closed then index $\gamma = m^{\gamma}_{L_0 L_1}$ of Definition 4.3.1, for $L_0 = \{\text{planes } y^i = \text{const. along } M\}$, $L_1 = TM$, in the symplectic vector bundle $E = T\mathbb{R}^{2n}|_M$. It is interesting to mention that the *Morse index* of the trajectories in configuration space of a dynamical system with Hamiltonian function H in $\mathbb{R}^{2n}$ (see Section 1.3) is equal to the Maslov index of an associated curve on a Lagrangian submanifold of $\mathbb{R}^{2n+2}$ [2]. It is the Maslov index as defined above which enters in the asymptotic solutions of the equations of quantum physics, and in the so-called *quantization conditions* (see later on an expression of these conditions). An important interpretation of the Maslov index as the index of a differential operator has been given recently by M. Gromov [36].

Among others, Arnold's proofs of [2] use the following interpretation of the class $m(L_0, L_1)$ defined by (4.3.5) for the particular L_0 and L_1 associated above with the Lagrangian submanifold $M \subset \mathbb{R}^{2n}$. Let $L(\mathbb{R}^{2n})$ be the manifold of the Lagrangian subspaces of $\mathbb{R}^{2n}$, which by Theorem 2.3.9 is isomorphic to $U(n)/O(n)$. Then the connecting mapping

$A(L_0,L_1)$ of 4.3.3 is in fact the mapping

$$\psi: M \to L(\mathbb{R}^{2n}) \approx U(n)/O(n) \qquad (4.3.12)$$

defined by $\psi(m) = T_m M$ $(m \in M)$. The cohomology group $H^1(U(n)/O(n), \mathbb{R}) = \mathbb{R}$, and it has the generator $g = (\det^2)^*[dz/2\pi\sqrt{-1}\,z]$, where

$$\det^2 : U(n)/O(n) \to S^1 \qquad (4.3.13)$$

is the square of the determinant of unitary matrices [37]. Hence (4.3.5) shows that $m(L_0,L_1)$ is the image of the generator g of $H^1(U(n)/O(n), \mathbb{R})$ by ψ^*.

It is interesting to notice that the class $\psi^*(g)$ can be also discovered by a straightforward analysis of the structure of a Lagrangian submanifold $M \subset \mathbb{R}^{2n}$. Let us remember the canonical atlas of such a manifold M, as provided by Proposition 3.3.1 and Corollary 3.3.2, and let us use again the notation encountered in that Proposition and Corollary. Let U_I, U_J $(I,J \subset \{1,2,\dots,n\})$ be two coordinate neighbourhoods of the canonical atlas of Corollary 3.3.2, and let us denote

$$I_1 = I \cap J,\ I_2 = I \setminus J,\ I_3 = J \setminus I,\ I_4 = \{1,\dots,n\} \setminus (I \cup J)\ . \qquad (4.3.14)$$

Then we have

$$I = I_1 \cup I_2,\ \bar{I} = I_3 \cup I_4,\ J = I_1 \cup I_3,\ \bar{J} = I_2 \cup I_4\ , \qquad (4.3.15)$$

the canonical coordinates on U_I are $(x^{I_1}, x^{I_2}, \xi_{I_3}, \xi_{I_4})$, the canonical coordinates on U_J are $(x^{I_1}, x^{I_3}, \xi_{I_2}, \xi_{I_4})$, and the Jacobian matrix of the transition relations between these two systems of local coordinates has the form

$$(4.3.16) \qquad \begin{pmatrix} Id & 0 & 0 & 0 \\ \frac{\partial x^{I_2}}{\partial x^{I_1}} & \frac{\partial x^{I_2}}{\partial x^{I_3}} & \frac{\partial x^{I_2}}{\partial \xi_{I_2}} & \frac{\partial x^{I_2}}{\partial \xi_{I_4}} \\ \frac{\partial \xi_{I_3}}{\partial x^{I_1}} & \frac{\partial \xi_{I_3}}{\partial x^{I_3}} & \frac{\partial \xi_{I_3}}{\partial \xi_{I_2}} & \frac{\partial \xi_{I_3}}{\partial \xi_{I_4}} \\ 0 & 0 & 0 & Id \end{pmatrix}$$

the notation being like in Proposition 3.3.1. The properties of the blocks of this matrix are given by Formula (3.3.2), and the result is that, over every intersection $U_I \cap U_J$, we have the nondegenerate symmetric matrix

$$(4.3.17) \qquad \mathfrak{M}_{IJ}(M) = \begin{pmatrix} \frac{\partial x^{I_2}}{\partial \xi_{I_2}} & \frac{\partial x^{I_2}}{\partial x^{I_3}} \\ -\frac{\partial \xi_{I_3}}{\partial \xi^{I_2}} & -\frac{\partial \xi_{I_3}}{\partial x^{I_3}} \end{pmatrix}$$

and the corresponding integers

$$(4.3.18) \qquad c_{IJ}(M) = \operatorname{sgn} \mathfrak{M}_{IJ}(M) \ ,$$

where sgn denotes the signature of the quadratic form associated to the respective matrix. Because of continuity, c_{IJ} are locally constant integers, and they define an integral 1-cochain of the covering $\{U_I\}$ in the sense of the Čech cohomology theory [75].

On the other hand, a proof similar to that of Proposition 3.3.1 shows that, if $L \in L(\mathbb{R}^{2n})$, L projects isomorphically onto at least one coordinate Lagrangian plane $x^{\bar{I}} = 0$, $\xi_I = 0$. Hence $L(\mathbb{R}^{2n})$ is covered by the open neighbourhoods

$$(4.3.19) \qquad W_I = \{L \in L(\mathbb{R}^{2n}) / L \text{ is isomorphic to } x^{\bar{I}} = 0, \xi_I = 0\} \ ,$$

and the same formula (4.3.17) yields the matrices $\mathfrak{A}_{JJ}(L(\mathbb{R}^{2n}))$, and the cochain

(4.3.20) $$c_{JJ}(L(\mathbb{R}^{2n})) = \operatorname{sgn} \mathfrak{A}_{IJ}(L(\mathbb{R}^{2n})) .$$

Now, the basic fact to be proven is that $c_{IJ}(L(\mathbb{R}^{2n}))$ is a 1-cocycle which defines the cohomology class 4g, g being the already considered generator of $H^1(L(\mathbb{R}^{2n}),\mathbb{R})$. Details about this construction, and the proof of the mentioned basic fact can be found in [61]. The class defined by (4.3.20) induces a cohomology class in M since (4.3.20) induces (4.3.18) on M via the set of the tangent spaces of M. The cohomology class μ obtained is 1/4 of the Maslov class (4.3.5), and it is involved in the so-called *quantization conditions*: a) the pull-back of the Liouville form λ of $\mathbb{R}^{2n}$ (formula (1.1.4)) to M is an exact form; b) $\mu \equiv 0 \pmod 8$, conditions which are important for theoretical physics applications [61], [58].

In order to indicate other approaches to the Maslov class, we introduce the *index*

(4.3.21) $$i(A,B,C) = \frac{1}{2} \operatorname{sgn} Q_A(B,C)$$

defined for three Lagrangian subspaces A,B,C of $\mathbb{R}^{2n}$ such that (A,B) and (A,C) are transversal pairs. In formula (4.3.21), $Q_A(B,C)$ is the quadratic form that defines the *vector* $\vec{BC}$ of the affine space $\mathcal{T}(A)$ of Theorem 2.2.15. This form is nondegenerate iff (B,C) is also a transversal pair of Lagrangian subspaces, hence it is natural to use it in the discussion of the transversality problem.

Let A' be another Lagrangian subspace which is also transversal to B and to C. Since $\mathcal{T}(B)$ is an affine space (Theorem 2.2.15) there is a line segment $AA'_{\mathcal{T}(B)}$. Similarly, there is a line segment $A'A_{\mathcal{T}(C)}$, and, together, these two segments yield a closed path γ in

$L(\mathbb{R}^{2n})$. The following fundamental formula can now be established (see a proof in [37])

$$(4.3.22)\qquad i(A,B,C) - i(A',B,C) = \int_{\gamma} (\det{}^2)^* \frac{dz}{2\pi\sqrt{-1}z} \quad ,$$

where $z \in S^1$, and $\det^2$ is the square of the determinant of the unitary matrices along γ in the identification $L(\mathbb{R}^{2n}) \approx U(n)/O(n)$. As a consequence, if A,A',B,C vary continuously, and the imposed transversality conditions remain satisfied, the second member of (4.3.22) is also continuous, and since the first member is an integer it is locally constant.

Now, let us consider the symplectic vector bundle $(\pi: E \to M,\omega)$ with the Lagrangian subbundles L_0, L_1. Let $\{U_\alpha\}_{\alpha\in A}$ be an open covering of M by trivializing neighbourhoods of E. In view of Proposition 2.2.17, it is clear that we have local Lagrangian subbundles $L_\alpha \subset E|_{U_\alpha}$ such that L_α is transversal to both L_0 and L_1, and, accordingly, we have the system of functions

$$(4.3.23)\qquad i_\alpha(L_0,L_1) = i(L_\alpha,L_0,L_1) : U_\alpha \to \mathbb{Z} .$$

If L_0, L_1 are transversal, then, by continuity reasons, $i_\alpha = \text{const.}$ but this is not true generally since the eigenvalues of the respective quadratic forms may have zeroes.

On the other hand, following Hörmander [41], we may also define the functions

$$(4.3.24)\qquad i_{\alpha\beta}(L_0,L_1) = i_\alpha(L_0,L_1) - i_\beta(L_0,L_1) : U_\alpha \cap U_\beta \to \mathbb{Z} .$$

As explained above, $i_{\alpha\beta}(L_0,L_1)$ are locally constant. Hence they define an integral onedimensional cocycle of the covering $\{U_\alpha\}$ in the sense of Čech cohomology. One can prove by using (4.3.22) that the corresponding class of $H^1(M,\mathbb{R})$ is just the Maslov class $m(L_0,L_1)$ of (4.3.5) [37]. We see again that $m(L_0,L_1) = 0$ for transversal subbundles L_0,L_1 since then $i_\alpha(L_0,L_1)$ = const. define a 1-cocycle, and by (4.3.24) $i_{\alpha\beta}(L_0,L_1)$ is the coboundary of this cocycle.

The index $i(A,B,C)$ of formula (4.3.21) was also studied by J. Leroy [51]. For Leray's theory it is basic to go over to the universal covering space of $L(\mathbb{R}^{2n})$. By looking at the point $1 \in S^1$, it is easy to see that the mapping $\det^2$ of (4.3.13) is a locally trivial fibre bundle with the simply connected fibre $SU(n)/SO(n)$ (S stands for the *special* subgroups of matrices with determinant = 1). The homotopy exact sequence of this fibration [75] yields

$$\text{(4.3.25)} \qquad \pi_1\left(L(\mathbb{R}^{2n})\right) = \pi_1\left(U(n)/O(n)\right) = \mathbb{Z} .$$

This result suggest that the universal covering space $\tilde{L}(\mathbb{R}^{2n})$ of $L(\mathbb{R}^{2n})$ should consist of pairs (X,θ) where $X \in \mathbb{R}^{2n}$, and θ is a *phase angle*, and it can indeed be checked that [37], [74]

$$\text{(4.3.26)} \qquad \tilde{L}(\mathbb{R}^{2n}) = \left\{(X,\theta) \mid X \in L(\mathbb{R}^{2n}),\ e^{i\theta} = \det \sigma(X)\right\} ,$$

where $\sigma\colon L(\mathbb{R}^{2n}) \to U(n)$ is the Souriau mapping defined in Section 2.3.

Now, for any three mutually transversal Lagrangian subspaces A,B,C, Leray gives a "coboundary" formula

$$\text{(4.3.27)} \qquad i(A,B,C) = m(\tilde{A},\tilde{B}) + m(\tilde{B},\tilde{C}) + m(\tilde{C},\tilde{A}) ,$$

where $\tilde{A},\tilde{B},\tilde{C}$ $\tilde{L}(\mathbb{R}^{2n})$, they project to A,B,C, respectively, and m is the *Leray index* which, by a formula of Souriau [74] is

$$\text{(4.3.28)} \qquad m(\tilde{A},\tilde{B}) = \frac{1}{2\pi}\left\{\theta_{\tilde{A}}-\theta_{\tilde{B}} + i \operatorname{tr} \operatorname{Log}\left(-\sigma(A)\sigma(B)^{-1}\right)\right\} .$$

In (4.3.28) the logarithm of a matrix $P \in G\ell(n,\mathbb{C})$ is defined for the case where P has no eigenvalues on the negative real axis, and it is given by

$$\text{(4.3.29)} \qquad \operatorname{Log} P = \int_{-\infty}^{0} \left\{(tI-P)^{-1} - (t-1)^{-1} I\right\} dt \qquad .$$

Furthermore, let M be a Lagrangian submanifold of $\mathbb{R}^{2n}$ with the associated Lagrangian subbundles $L_0 = \{y^i=\text{const}\}|_M$, $L_1 = TM$. Let $\gamma: [0,1] \to M$ be a closed curve at $x_0 \in M$ such that $T_{x_0}M$ is transversal to the plane y^i=const., and let $\tilde{\gamma}$ be a lift of the curve $T_\gamma M$ of $L(\mathbb{R}^{2n})$ to $\tilde{L}(\mathbb{R}^{2n})$. Then, the Maslov index of γ is given by

$$\text{index}\,\gamma = m(\tilde{X}_0, \tilde{\gamma}(1)) - m(\tilde{X}_0, \tilde{\gamma}(0)) , \tag{4.3.30}$$

where $\tilde{X}_0 \in \tilde{L}(\mathbb{R}^{2n})$ projects to the plane y^i=const.

The literature on Maslov's class contains also generalizations to higher cohomological dimensions. First, such a generalization has been defined by V.I. Arnold and studied by D.B. Fuks [29]. More recently, for submanifolds of cotangent bundles, a generalization of the Maslov class has been given by V.A. Vasiliev [85], [6]. The classes obtained thereby are important for the theory of Lagrangian cobordism [4]. Directly related to the transversality problem is another recent generalization of the Maslov class due to Morvan [63] and Morvan and Niglio [66]. This later generalization is obtained for the case of trivial symplectic vector bundles. An interesting general construction which uses stable vector bundles and a review of various definitions of higher order Maslov classes is given by M. Audin [7]. We do not give details about all these higher order Maslov classes since we do not intend to compare them with the secondary characteristic classes of the present work. The interested reader is referred to the literature quoted above, and he will discover that it is natural to expect strong relations between the secondary characteristic classes of the next section and the classes of Fuks-Vasiliev-Audin-Morvan-Niglio, but these relations are not yet clear. In any case, the secondary characteristic classes considered here are naturally associated with general Lagrangian subbundles and they provide both transversality obstructions, and numerical invariants of cylindrical Lagrangian cobordism (see later on Sections 4.4, 4.5). At the same time, these secondary characteristic classes include Maslov's class, which is the most important for applications.

In fact, it has been known since Arnold's paper [2] that the Maslov class is a characteristic class in the category of real vector bundles with a trivial complexification. (In the case of a Lagrangian submanifold $M \subset \mathbb{R}^{2n}$, $T\mathbb{R}^{2n}/_M \approx T^cM = TM \otimes_{\mathbb{R}} \mathbb{C}$.) This remark is made precise by Kamber and Tondeur [43], who determined the secondary characteristic classes of trivial complex vector bundles with a real reduction (i.e., a reduction of the structure group from $G\ell(n,\mathbb{C})$ to $G\ell(n,\mathbb{R})$), and proved that the first secondary class is $-1/2$ of the Maslov class. Hence the other secondary classes of this configuration can be seen as higher order Maslov classes.

The remark of Kamber and Tondeur was the starting point of the present work. Indeed, if we think of secondary characteristic classes as defined in Section 4.2, and if we consider the (not necessarily trivial) symplectic vector bundle E with the Lagrangian subbundles L_0, L_1, then we notice the existence of two different reductions of the structure group of E to $O(n)$, and, accordingly, we notice the existence of two different families of orthogonal connections on E. These families yield secondary characteristic classes. It is easy to understand that if L_0 and L_1 are transversal subbundles, the two families of orthogonal connections have a non void intersection, and the secondary classes vanish. Hence, these classes are transversality obstructions. It is precisely this theory that we want to develop in details, and this will be the object of our next section.

4.4. MASLOV SECONDARY CHARACTERISTIC CLASSES. In this section, we shall study the characteristic classes of symplectic vector bundles and Lagrangian subbundles.

Let $\pi: E \to M^m$ be a symplectic vector bundle of rank $2n$, and with the symplectic form ω. It is known from Theorem 3.1.2 that this bundle admits a unique "homotopy" class of compatible positive complex structures or, equivalently, reductions of the structure group from $Sp(n,\mathbb{R})$ to $U(n)$. Let J be one such structure, fixed for the moment, and let $\mathcal{U}(E,J)$ be the corresponding bundle of complex unitary frames of E, as defined by formula (3.1.4). $\mathcal{U}(E,J)$ is a $U(n)$-principal bundle over M, and it has primary characteristic classes associated to the polynomials of the invariant Weil algebra $I(U(n))$.

The algebra $I(U(n))$ is well known (see for instance [45]), and we describe again its structure here. One has

Theorem 4.4.1. *Let* $A \in \mathfrak{u}(n)$, *and*

$$c_k(A) = \left(\frac{-1}{2\pi\sqrt{-1}}\right)^k \operatorname{tr} \wedge^k A \qquad (k = 1,2,\ldots,n) \ , \tag{4.4.1}$$

where $\wedge^k A$ *is the matrix of the* k^{th} *order minors of* A. *Then*

$$I(U(n)) = \mathbb{R}[c_1,\ldots,c_n] \ . \tag{4.4.2}$$

Proof. The polynomials (4.4.1) appear as the coefficients of the development

$$\det\left(\lambda I_n - \frac{1}{2\pi\sqrt{-1}} A\right) = \sum_{k=0}^{n} c_k(A)\lambda^{n-k} \qquad (c_0(A) = 1) \ , \tag{4.4.3}$$

where I_n is the nth order unit matrix, and λ is an indeterminate. Since $A \in \mathfrak{u}(n)$ means

$$A + {}^t\bar{A} = 0 \ , \tag{4.4.4}$$

it follows that the left-hand side of (4.4.3) is real, and so are the coefficients $c_k(A)$. Moreover, it is classical linear algebra that the determinant (4.4.3) is invariant by $\operatorname{ad} T$ for any $T \in G\ell(n,\mathbb{C})$. Hence $c_k \in I^k\{u(n)\}$.

Now, if $A' = (1/2\pi\sqrt{-1})A$, (4.4.4) yields $A' = {}^t\bar{A}'$, i.e., A' is a hermitian matrix. Again, it is known from linear algebra, that one can choose $T \in G\ell(n,\mathbb{C})$ such that $T^{-1}A'T$ takes the diagonal form

$$(4.4.5) \qquad \begin{pmatrix} a_1 & & 0 \\ & & \\ 0 & & a_n \end{pmatrix} \qquad (a_i \in \mathbb{R})$$

We see that, because of the previous invariance property, c_k are proportional to the basic symmetric polynomials in $a_1,\ldots,a_n$ [46]. On the other hand, for every $f \in I^k\{U(n)\}$, we obtain in the same manner some symmetric polynomial in $a_1,\ldots,a_n$, and it is known that the latter will be a polynomial in $c_1,\ldots,c_n$ [46]. Formula (4.4.2) is now clear. Q.e.d.

Definition 4.4.2. *The polynomial* c_k *of (4.4.1) is called the* k^{th} ***Chern polynomial.***

The explicit expression of c_k is

$$(4.4.6) \qquad c_k(A) = \frac{(-1)^k}{(2\pi\sqrt{-1})^k k!}\,\delta^{j_1\cdots j_k}_{i_1\cdots i_k}\, a^{i_1}_{j_1} \cdots a^{i_k}_{j_k},$$

where $\delta^{\cdots}_{\cdots}$ is the well known Kronecker multiindex.

A simple but basic fact for our purposes is

Proposition 4.4.3. *All the Chern polynomials of odd order are identically zero on the orthogonal subalgebra* $o(n) \subset u(n)$.

Proof. If $A \in o(n)$, then (4.4.4) becomes $A + {}^tA = 0$, and it is well known that a skew symmetric determinant is zero. Hence $c_{2h-1} = 0$ for $h = 1,2,\ldots$. Q.e.d.

Remark 4.4.4. *The same proof as for Theorem 4.4.1, but using*

$$\det\left(\lambda I_n - \frac{1}{2\pi} B\right), \quad B \in o(n) \quad ,$$

yields that the invariant Weil algebra of the orthogonal group $O(n)$ *is*

$$(4.4.7) \qquad I\big(O(n)\big) = \mathbb{R}\,[p_1,\ldots,p_{[\frac{n}{2}]}] \quad ,$$

where

$$(4.4.8) \qquad p_j(B) = \left(\frac{1}{2\pi}\right)^{2j} \operatorname{tr} \wedge^{2j} B =$$

$$= \frac{1}{(2\pi)^{2j}\,(2j)!}\, \delta^{k_1\ldots k_{2j}}_{h_1\ldots h_{2j}}\, b^{h_1}_{k_1} \cdots b^{h_{2j}}_{k_{2j}} \quad .$$

The similar polynomials for $\wedge^{2j+1} B$ *are zero by the same reason as in Proposition 4.4.3.)* p_j *are called the* ***Pontrjagin polynomials****, and we see that*

$$(4.4.9) \qquad p_j = (-1)^j c_{2j} /_{o(n)} \quad .$$

Finally, let us indicate that the factor $1/2\pi$ *is introduced in order to obtain from the Chern and the Pontrjagin polynomials characteristic classes which are real images of integral classes* [45].

Definition 4.4.5. *If* $P \to M$ *is a* $U(n)$*-principal bundle, the characteristic classes* $c_k(P) = w\big(U(n),P\big)c_k \in H^{2k}(M,\mathbb{R})$, *where* w *is the Chern-Weil homomorphism (formula (4.2.19)) are called the Chern classes of* P. *Particularly* $e(P) = c_n(P)$ *is the* ***Euler class*** *of* P *(see* [45] *for an independent definition of the Euler class). If* $P \to M$ *is an* $O(n)$*-principal bundle then* $p_j(P) = w\big(O(n),P\big)p_j \in H^{4j}(M,\mathbb{R})$ *are called the* ***Pontrjagin classes*** *of* P.

Proposition 4.4.6. *If the* U(n)*-principal bundle* $P \to M$ *has an* O(n)*-reduction* P', *then* $c_{2h-1}(P) = 0$, *and* $p_h(P') = (-1)^h c_{2h}(P)$.

Proof. This is an immediate consequence of Proposition 4.4.3 and formula (4.4.9). Q.e.d.

Proposition 4.4.7. *The Chern and Pontrjagin classes are real images of integral Chern and Pontrjagin classes, and*

$$c_k \in I_0\{U(n)\}, \quad p_j \in I_0\{O(n)\} \quad , \tag{4.4.10}$$

where I *is the subalgebra defined by formula (4.2.45).*

Proof. As recalled in Section 4.1 the classifying spaces $B_{U(n)}$ and $B_{O(n)}$ are complex and real Grassmann manifolds, and the computation of the cohomology of these manifolds shows that the corresponding universal Chern-Weil homomorphisms send c_k and p_k, respectively, to integral cohomology classes. The actual computation and proof would take us too far away from our subject. It can be found, for instance, in [60]. Q.e.d.

Now, let us come back to the symplectic vector bundle E of the beginning of this Section, and to the associated U(n)-principal bundle $U(E,J)$. We can prove

Proposition 4.4.8. *If* J_0 *and* J_1 *are two compatible positive complex structures of* E *then*

$$c_k\{U(E,J_0)\} = c_k\{U(E,J_1)\} \ . \tag{4.4.11}$$

Proof. Let J_s, $s \in I = [0,1]$, be a differentiable family of compatible positive complex structures such that J_0 and J_1 are the two given structures (Theorem 3.1.2), and let g_s be the corresponding family of hermitian metrics. Then, we can see that

$$(4.4.12) \qquad \pi\colon \mathcal{U}(J_0,J_1) \overset{\text{def}}{=} \bigcup_{x\in M,\, s\in[0,1]} \mathcal{U}(E,J_s)_x \to M\times I\ ,$$

where the index x denotes the fibre at x, and π is the natural projection, is a $U(n)$-principal differentiable bundle over $M\times I$. Indeed, for every $(x,s)\in M\times I$ we can find an open neighbourhood $V_{(x,s)}$ on which we can construct a field of unitary frames of the metrics g_s by means of the classical orthonormalization procedure. In this orthonormalization procedure, the dependence on the paramater s is clearly differentiable. Thereby, we obtain local cross sections $\sigma_{V_{(x,s)}}$ of $\mathcal{U}(J_0,J_1)$, and, as usual in fiber bundles, this yields bijections

$$(4.4.13) \qquad \pi^{-1}(V_{(x,s)}) \to V_{(x,s)} \times U(n)\ .$$

By asking the bijections (4.4.13) to be diffeomorphisms, we obtain the indicated bundle structure of $\mathcal{U}(J_0,J_1)$.

Now, let us consider the differential form $\Delta(\chi)c_k$, where χ is a $U(n)$-connection on $\mathcal{U}(J_0,J_1)$, and $\Delta(\chi)$ is the Chern-Weil homomorphism (4.2.17). This form is on $M\times I$, and we may use for it the Stokes formula (4.1.29). The result is

$$(4.4.14) \qquad d\int_I \Delta(\chi)c_k - \int_I d\Delta(\chi)c_k = \Delta(\chi_1)c_k - \Delta(\chi_0)c_k\ ,$$

where χ_0 and χ_1 are the unitary connections induced by χ in $\mathcal{U}(E,J_0)$, $\mathcal{U}(E,J_1)$, respectively. Since $\Delta(\chi)c_k$ is closed, (4.4.14) obviously implies (4.4.11). Q.e.d.

In view of Proposition 4.4.8 it is meaningful to define

Definition 4.4.9. *The Chern classes* $c_k\big(\mathcal{U}(E,J)\big)$ *(which do not depend on the choice of* J*) are called the* ***Chern classes*** $c_k(E)$ ***of the symplectic vector bundle*** E. *Particularly,* $c_n(E)$ *is the* ***Euler class*** *of* $e(E)$.

If E=TM, *where* M *is an almost symplectic manifold the classes defined above are called the* ***Chern and Euler classes*** *of* M, *respectively.*

For the completeness of the formulas let us write down the expression of the differential forms which define the Chern classes. Let θ be a unitary connection on $\mathcal{U}(E,J)$. Like for any linear connection, we may write local connection equations (4.1.16). More precisely, we cover M by trivializing neighbourhoods of E, and take fields of unitary frames (e_i, Je_i) $(i = 1,\dots,n)$ over these neighbourhoods. In the complex form, we have the frames of formula (2.3.13), i.e.,

$$\varepsilon_i = \frac{1}{\sqrt{2}}(e_i - \sqrt{-1}\,Je_i) \quad , \tag{4.4.15}$$

and the local equations of θ are of the form

$$\nabla\varepsilon_i = \theta_i^j\varepsilon_j \quad , \tag{4.4.16}$$

where the matrix (θ_i^j) takes values in $u(n)$. The corresponding local curvature forms Θ_i^j are given by (4.1.21), and from (4.2.14), (4.4.6) we get

$$\Delta(\theta)c_k = \frac{(-1)^k}{(2\pi\sqrt{-1})^k\,k!}\,\delta^{j_1\dots j_k}_{i_1\dots i_k}\,\Theta^{i_1}_{j_1}\wedge\dots\wedge\Theta^{i_k}_{j_k} \quad . \tag{4.4.17}$$

These are the representative forms of the Chern classes of E.

Remark 4.4.10. *Of course, because of (4.4.2), the conclusion of Proposition 4.4.8 holds good for all of the primary characteristic classes of* $\mathcal{U}(E,J)$, *hence we may consider the whole algebra* $w\big(U(n),E\big)\big(I(U(n))\big)$ *of characteristic classes of the symplectic vector bundle* E.

Now, we shall study the situation where the symplectic vector bundle $E \to M$ above has a Lagrangian subbundle L_0. We begin with the following simple remarks [87], [23].

Proposition 4.4.11. *If the symplectic vector bundle* E *admits a Lagrangian subbundle* L_0 *then the odd dimensional Chern classes of* E *are zero.*

Proof. By Proposition 3.1.6, we know that the existence of L_0 implies the reducibility of the structure group of E to O(n), hence $c_{2h-1}(E) = 0$ follows from Proposition 4.4.6. Q.e.d.

Remark 4.4.12. *The Euler class* e(E) *vanishes if* n *is odd, and it is a "square" if* n=2h. *Indeed, by a known property of skew symmetric determinants of an even order (see, for instance, [80], p. 243, formula (25)), we have*

$$(4.4.18) \qquad \Delta(\theta)c_{2h} = \frac{(-1)^{2h}}{(2\pi\sqrt{-1})^{2h}} \det \Theta =$$

$$= \frac{(-1)^{h}}{(2\pi)^{2h}} \frac{1}{(2^h h!)^2} \left[\sum_{\sigma} \varepsilon_{i_1 \dots i_{2h}} \Theta^{i_1}_{i_2} \wedge \dots \wedge \Theta^{i_{2h-1}}_{i_{2h}}\right]^2 ,$$

where Θ *is the curvature of an* O(n)*-connection on* L_0 , σ *runs over all the permutations* $(i_1,\dots,i_{2h})$ *of* $(1,\dots,2h)$, *and* $\varepsilon_{i_1 \dots i_{2h}}$ *is the sign of the permutation* $(i_1,\dots,i_{2h})$. *If* L_0 *is orientable it has also a Euler class [45], and formula (4.4.18) means*

$$(4.4.19) \qquad e(E) = (-1)^h [e(L_0)]^2 .$$

Remark 4.4.13. *Proposition 4.4.11 allows us to give some more examples of symplectic vector bundles which do not have Lagrangian subbundles (see also Section 3.1). For instance, the Chern classes of the* n*-dimensional complex projective space* $\mathbb{C}P^n$ *are nonzero (e.g., [16]), hence, in view of Proposition 4.4.11,* $T\mathbb{C}P^n$ *has no Lagrangian subbundles.*

Let us consider again the symplectic bundle $\pi : E \to M$ with the Lagrangian subbundle L_0, with the positive complex structure J, and the hermitian metric g. Let $U(E,J,L_0)$ be the O(n)-principal subbundle of the L_0-related unitary frames, defined by (3.1.6), and which gives the reduction of the structure group of $\mathcal{U}(E,J)$ from $U(n)$ to $O(n)$. We shall denote by θ_0 a connection on $\mathcal{U}(E,J,L_0)$, and, at the same time, its extenstion to $\mathcal{U}(E,J)$ (Section 4.1). The connection θ_0 has local equations of the form (4.4.16), where (ε_i) is a local field of L_0-related unitary bases associated to the g-orthonormal bases (e_i) of L_0, and, accordingly, the relations (4.1.23) hold good.

Definition 4.4.14. *A connection* θ_0 *on* $\mathcal{U}(E,J)$ *of the type described above will be called an* L_0***-orthogonal unitary connection.***

If θ_0 is the connection described above, then, by Proposition 4.4.3, we have

$$c_{2h-1} \in \ker \Delta(\theta_0) \qquad (h = 1,2,\ldots) \,, \tag{4.4.20}$$

and, in view of Definition 4.2.9 we may give now

Definition 4.4.15. *The Chern-Simons classes*

$$\mu^h(E,J;L_0) = [T(\theta_0)c_{2h-1}] \in H^{4h-3}\big(\mathcal{U}(E,J),\mathbb{R}\big) \tag{4.4.21}$$

will be called ***Chern-Simons-Maslov classes*** *of* (E,L_0) *with respect to* J.

The forms $T(\theta_0)c_{2h-1}$ are defined by formula (4.2.29). In our case, the local connection forms are θ_i^j of (4.4.16), and, therefore, the connection form θ of (4.2.29), is given by (4.1.19), i.e.,

$$\theta^j_{oi} = \eta^j_h\xi^k_i\theta^h_k + \eta^j_h d\xi^h_i \quad , \tag{4.4.22}$$

where ξ is an arbitrary unitary matrix, and $\eta = \xi^{-1}$. Furthermore, the curvature forms of θ_0 are, by (4.1.20), equal to

$$\Theta^j_{oi} = \eta^j_h\xi^k_i\Theta^h_k \quad , \tag{4.4.23}$$

and then, in view of (4.2.28), the curvature $\bar{\Theta}_t$ of (4.2.29) is given by

$$(4.4.24) \qquad (\bar{\theta}_t)^j_i = t\theta^j_{oi} + t(1-t)\theta^h_{oi} \wedge \theta^j_{oh} =$$

$$= t\theta^j_{oi} + t(1+t)d\theta^j_{oi} \quad .$$

Now, if we use the formulas (4.4.6) and (4.2.6), we can finally write down (4.2.29) in our case as

$$(4.4.25) \qquad T(\theta_0)c_{2h-1} = (-1)^{h+1} \frac{\sqrt{-1}}{(2\pi)^{2h-1}(2h-2)!} \cdot$$

$$\cdot \int_0^1 \left\{ \delta^{j_1 \ldots j_{2h-1}}_{i_1 \ldots i_{2h-1}} \theta^{i_1}_{o\,j_1} \wedge (\bar{\theta}_t)^{i_2}_{j_2} \wedge \ldots \wedge (\bar{\theta}_t)^{i_{2h-1}}_{j_{2h-1}} \right\} dt.$$

Remark 4.4.16. *Since the matrices* θ_0 *and* $\bar{\theta}_t$ *are skew symmetric with respect to* L_0*-related unitary bases, it follows that the pullback of* $T(\theta_0)c_{2h-1}$ *to* $\mathcal{U}(E,J,L_0)$ *is zero, which means that, in fact,* $\mu^h(E,J,L_0)$ *can be seen as an element of the relative cohomology space* $H^{4h-3}(\mathcal{U}(E,J),\mathcal{U}(E,J,L_0);\mathbb{R})$, *but we are not interested in this fact. Let us notice that one can consider classes (4.4.21) in* $H^{4h-3}(\mathcal{B}(E),\mathbb{R})$, *where* $\mathcal{B}(E)$ *is the principal bundle of all the linear frames of* E, *as soon as the later has an* $O(n)$*-reduction* $\mathcal{O}(E)$. *But these classes are zero since* $\mathcal{B}(E)$ *retracts onto* $\mathcal{O}(E)$, *and the forms* $T(\theta_0)c_{2h-1}$ *vanish on* $\mathcal{O}(E)$ [18]. *On the contrary, there is no retraction of* $\mathcal{U}(E,J)$ *onto* $\mathcal{U}(E,J,L_0)$, *which makes the Chern-Simons-Maslov classes meaningful.*

Now, let us establish some basic properties of these classes.

Proposition 4.4.17. *The classes* $\mu^h(E,J;L_0)$ *do not depend on the choice of the connection* θ_0 *on* $\mathcal{U}(E,J,L_0)$.

Proof. Let θ_0' be another L_0-orthogonal unitary connection. Then $(1-s)\theta_0 + s\theta_0'$, $s \in I = [0,1]$, is a link between θ_0 and θ_0', which is the extension to $U(E,J) \times I$ of an $O(n)$-connection on $L_0 \times I \to M \times I$. This proves that θ_0 and θ_0' are c_{2h-1}-homotopic connections (Definition 4.2.15), and the conclusion follows from Theorem 4.2.16. Q.e.d.

Proposition 4.4.18. *The classes* $\mu^h(E,J;L_0)$ *are rigid (i.e., remain unchanged) with respect to homotopic deformations of* L_0 *via Lagrangian subbundles of* E.

Proof. Let L_s, $s \in I = [0,1]$, be a differentiable family of Lagrangian subbundles of E, starting with the bundle L_0 of the proposition. Then, by the same reason as in Proposition 4.4.8, we see that

$$V(L_s) = \bigcup_{x \in M, s \in I} U(E,J;L_s)_x \to M \times I \tag{4.4.26}$$

is a $O(n)$-principal bundle which defines a reduction of the $U(n)$-bundle $U(E,I) \times I \to M \times I$. Hence, there exists a connection χ on $U(E,J) \times I$ which extends a connection on $V(L_s)$ and, therefore, χ is a c_{2h-1}-homotopy between $\chi|_{s=0}$ and $\chi|_{s=1}$. Since in view of Proposition 4.4.17 we may compute $\mu^h(E,J;L_a)$ by using precisely the connections $\chi|_{s=a}$ $(a=0,1)$, the conclusion follows again from Theorem 4.2.16. Q.e.d.

Remark 4.4.19. *If* J' *is another compatible positive complex structure, then* $\mu^h(E,J';L_0)$ *belong to a different space. They can be compared with* $\mu^h(E,J;L_0)$ *by using the bundle* $U(J_0,J_1)$ *of (4.4.12), where* $J_0 = J$, $J_1 = J'$. *Indeed, if for every* $s \in [0,1]$ *we consider the* L_0-*related unitary frames with respect to the metric* g_s, *we obtain an* $O(n)$-*reduction of* $U(J_0,J_1)$, *and we get Chern-Simons classes* $[T(\theta)c_{2h-1}] \in H^{4h-3}(U(J_0,J_1), \mathbb{R})$, *where* θ *is an orthogonal connection on* $U(J_0,J_1)$. *For* $s=0$, *these classes induce* $\mu^h(E,J;L_0)$, *and for* $s=1$

they induce $\mu^h(E,J';L_0)$. *Or, from another viewpoint, by using the cross-sections* $\sigma_{V_{(x,s)}}$ *of the proof of Proposition 4.4.8, and by dividing* $[0,1]$ *into a finite number of segments where sections* $\sigma_{V_{(x,s)}}$ *are defined, we can define a "translation" of unitary frames which establishes an isomorphism of the principal bundles* $\mathcal{U}(E,J)$, $\mathcal{U}(E.J')$. *The classes* $\mu^h(E,J;L_0)$, $\mu^h(E,J';L_0)$ *correspond to each other by this isomorphism. However, the considered isomorphism is not a canonical one.*

The simplest case of a Chern-Simons-Maslov class is for $h=1$. In this case, (4.4.25) yields

$$(4.4.27)\qquad T(\theta_0)c_1 = \frac{\sqrt{-1}}{2\pi}\theta^i_{oi} = \frac{\sqrt{-1}}{2\pi}\eta^i_h d\xi^h_i = \frac{\sqrt{-1}}{4\pi} d\,\ell n \det{}^2(\xi^j_i) \quad ,$$

where the second equality follows from (4.4.2) and (4.1.23), and the last equality follows by a known matrix computation.

Proposition 4.4.20. *The class* $\mu^1(E,J;L_0)$ *is induced by the generator of* $H^1(S^1)$ *via the mapping* $(-1/2)\det^2$ *and the bundle structure of* $\mathcal{U}(E,J)$.

Proof. Let U be a trivializing neighbourhood of $\mathcal{U}(E,J)$ with the trivializing diffeomorphism

$$(4.4.28)\qquad h_U : \pi^{-1}(U) \to U \times U(n) \quad ,$$

and let $p: U(n) \to U(n)/O(n)$ be the natural projection. Then formula (4.4.27) means precisely

$$(4.4.29)\qquad T(\theta_0)c_1 = -\frac{1}{2} h_U^* pr_{U(n)}^* p^*(\det{}^2)^* \left(\frac{dz}{2\pi\sqrt{-1}\,z}\right) \quad ,$$

where $z \in S^1 = \{z \in \mathbb{C}/|z|=1\}$. The result of (4.4.29) does not depend on the choice of h_U since the transition functions of $\mathcal{U}(E,J)$ may be taken in $O(n)$. Q.e.d.

Remark 4.4.21. *Formula (4.4.29) makes sense also if* h_U *are trivializing diffeomorphisms of the Lagrange Grassmannian bundle* $L(E)$ *of diagram (3.1.5). Hence, we have also a cohomology class*

$$\bar{\mu}^{1}(E,L_0) \in H^1(L(E),\mathbb{R}) , \tag{4.4.30}$$

and it is clear that

$$\mu^{1}(E,J;L_0) = \sigma^* \bar{\mu}^{1}(E,L_0) , \tag{4.4.31}$$

where $\sigma\colon U(E,J) \to L(E)$ *is defined by (3.1.5). It is easy to understand that* $\bar{\mu}^{1}(E,L_0)$ *does not depend on the choice of* J *because any two such structures* J *are homotopic. If* L_0 *is seen as a section* $\ell_0\colon M \to L(E)$, *then* $\ell_0^*(\bar{\mu}^{1}(E,L_0)) = 0$. *In view of the explanations of Section 4.3, formula (4.4.29) indicates a relation of* μ^1, *and the Maslov class, which explains the name used in Definition 4.4.15, and which will be clarified later on.*

Generally, we do not have results like Proposition 4.4.20 for μ^h with $h > 1$, and such classes μ^h also depend on the Pontrjagin classes of L_0. However, we can prove

Proposition 4.4.21. *If the Lagrangian subbundle* L_0 *admits a flat metric connection (and, particularly, if* L_0 *is trivial), the complex structure* J *can be chosen such that* $\mu^h(E,J;L_0)$ *will be induced by cohomology classes of* $U(n)/O(n)$.

Proof. The hypothesis means that L_0 has a Euclidean metric h_0 and a metric (i.e., orthogonal) connection θ_0 of zero curvature. Then, we shall take a transversal Lagrangian subbundle L_1 and J such that $L_1 = JL_0$, and extend connection θ_0 to $U(E,J)$. Since θ_0 has no curvature, the local frames (e_i) and (ε_i) of (4.4.16) may be taken to be parallel, i.e., $\theta_i^j = 0$ in (4.4.16). Now, the conclusion follows straightforwardly from formulas (4.4.22), (4.4.24), (4.4.25), and from the fact that the transition functions lf L_0 and of $U(E,J)$ are locally constant because of the zero curvature hypothesis. Q.e.d.

Remark 4.4.22. *In the situation of Proposition 4.4.21, we may use again diagram (3.1.5), and consider first the corresponding cohomology classes induced from* $U(n)/O(n)$ *to* $L(E)$, *then lift by* σ *to* $U(E,J)$. *The classes obtained on* $L(E)$ *were considered by Fuks (classes* b_k *of* [29], *and Morvan and Niglio* [66].

An example of the situation described in Proposition 4.4.21 can be obtained as follows. Let (V^{2n},ω) be an almost symplectic manifold, and let M be a Lagrangian submanifold of V such that M is a leaf with a compact linear holonomy group of a (not necessarily Lagrangian) foliation F of V [70]. Then, if we take $E = TV/_M$, $L_0 = TM$, L_0 has a flat metric connection. Indeed, $TM \approx (E/TM)^* = (TV/TF)^*/_M$, and the holonomy hypothesis shows that the structure group of TM is reducible to a compact linear group, while the transition functions of TM are locally constant functions (they are transition functions of a dual transversal bundle of a foliation, along a leaf). Since a compact linear group preserves a metric, we can obviously obtain a flat metric connection on our L_0. Lagrangian leaves of foliations of symplectic manifolds are discussed in [65].

Finally, let us prove

Proposition 4.4.23. *The cohomology classes* $2\mu^h(E,J;L_0)$ *are real images of integral cohomology classes.*

Proof. We follow, in fact, a similar proof of [18]. By the classification theorem of Narasimhan-Ramanan (Theorem 4.1.3), the connection θ_0 of L_0 used in (4.4.21) is induced by the universal $O(n)$-connection ν on a Stiefel bundle $\mathcal{V}(m,n)$ over a Grassmann manifold $\mathfrak{G}(m,n)$ with a large m. Hence the extension of θ_0 to $\mathcal{U}(E,J)$ is induced by the extension of ν to $\hat{\mathcal{V}}(m,n)$, where the latter is obtained from $\mathcal{V}(m,n)$ by the extension of the structure group to $U(n)$. I.e., we have a bundle morphism

$$\begin{array}{ccc} \mathcal{U}(E,J) & \overset{\varphi}{\to} & \hat{\mathcal{V}}(m,n) \\ \pi\downarrow & & \downarrow p \\ M & \overset{f}{\to} & \mathfrak{G}(m,n) \end{array} \tag{4.4.32}$$

such that $\theta_0 = \varphi^*\nu$, which clearly implies

$$T(\theta_0)c_{2h-1} = \varphi^* T(\nu)c_{2h-1} \quad . \tag{4.4.33}$$

Now, in view of Proposition 4.4.7, we may apply Proposition 4.2.18, which gives

$$[T(\theta_0)c_{2h-1}] = \varphi^*[T(\nu)c_{2h-1}] = \varphi^* p^* u \quad (\mathrm{mod}\ \mathbb{Z}) \ , \tag{4.4.34}$$

where $u \in H^{4h-3}(\mathfrak{G}(m,n), \mathbb{R}/\mathbb{Z})$.

But, it is a known cohomological property of the Grassmann manifold $\mathfrak{G}(m,n)$ that $2u = 0$ [14], [60]. Hence $2[T(\theta_0)c_{2h-1}] = 0$ (mod $\mathbb{Z}$). Q.e.d.

Now, we shall proceed and discuss the case of a symplectic vector bundle $\pi: E \to M$ endowed with two Lagrangian subbundles L_0 and L_1. We shall use the same notation as previously, and then we see that the $U(n)$-principal bundle $\mathcal{U}(E,J)$ has two $O(n)$-reductions $\mathcal{U}(E,J;L_a)$ $(a=0,1)$, and, if θ_a are L_a-orthogonal unitary connections on $\mathcal{U}(E,J)$, then $c_{2h-1} \in \ker \Delta(\theta_0) \cap \ker \Delta(\theta_1)$. Hence, in agreement with Definition 4.2.6 we may formulate

Definition 4.4.24. *The simple secondary characteristic classes*

$$\mu_h(E,L_0,L_1) = [\Delta(\theta_0,\theta_1)c_{2h-1}] \in H^{4h-3}(M,\mathbb{R}) \tag{4.4.35}$$

will be called the ***Maslov classes of the triple*** (E,L_0,L_1).

In order to write down a formula for $\Delta(\theta_0,\theta_1)c_{2h-1}$, we consider local fields of L_a-related unitary frames

$$\overset{a}{\varepsilon}_i = \frac{1}{\sqrt{2}}\left(\overset{a}{e}_i - \sqrt{-1}\, J\overset{a}{e}_i\right) \quad (\overset{a}{e}_i \in L_a;\ a=0,1)\ , \tag{4.4.36}$$

and consider L_a-orthogonal unitary connections θ_a given locally by

$$\nabla\overset{a}{\varepsilon}_i = \underset{a}{\omega}{}_i^j\,\overset{a}{\varepsilon}_j \qquad (\underset{a}{\omega}{}_i^j + \underset{a}{\omega}{}_j^i = 0;\ a=0,1). \tag{4.4.37}$$

(This time, we use ω instead of θ to avoid confusion between the local and the global connection forms.) Since the final expression has to be referred to common frames, let ε_i be some fixed local unitary frames, i.e., cross sections of $\mathcal{U}(E,J)$. Then we have some relations

$$\varepsilon_i = \overset{a}{\gamma}{}_i^j\,\overset{a}{\varepsilon}_j \quad \left((\overset{a}{\gamma}{}_i^j) \in U(n);\ a = 0,1\right)\ , \tag{4.4.38}$$

and, by (4.1.18), the new local connection forms of θ_a will be

$$\overset{a}{\pi}{}_i^j = \overset{a}{\nu}{}_h^j\, d\,\overset{a}{\gamma}{}_i^h + \overset{a}{\nu}{}_h^j\,\overset{a}{\gamma}{}_i^k\,\overset{a}{\omega}{}_k^h \qquad (a = 0,1) \tag{4.4.39}$$

where ν is the inverse matrix of γ. Now, we just have to write down (4.2.26) where $f = c_{2h-1}$ of (4.4.6), and the arguments may be taken to be the local forms

$$\alpha_i^j = \overset{1}{\pi}{}_i^j - \overset{0}{\pi}{}_i^j \tag{4.4.40}$$

$$\begin{aligned}(\theta_t)_i^j &= d\left(\overset{0}{\pi}{}_i^j + t\,\alpha_i^j\right) - \left(\overset{0}{\pi}{}_i^h + t\,\alpha_i^h\right) \wedge \\ &\wedge \left(\overset{0}{\pi}{}_h^j + t\,\alpha_h^j\right) \quad (\text{modulo } dt) = \\ &= (1-t)\overset{0}{\Omega}{}_i^j + t\,\overset{1}{\Omega}{}_i^j + t(1-t)\alpha_i^h \wedge \alpha_h^j\ ,\end{aligned} \tag{4.4.41}$$

where $\overset{a}{\Omega}{}_i^j$ are the local curvature forms of the connections θ_a, i.e.,

$$\overset{a}{\Omega}{}_i^j = d\overset{a}{\pi}{}_i^j - \overset{a}{\pi}{}_i^h \wedge \overset{a}{\pi}{}_h^j \qquad (a = 0,1) \tag{4.4.42}$$

(We may use the local forms because of the ad-invariance property of

c_{2h-1}.) The result is

(4.4.43) $$\Delta(\theta_0,\theta_1)c_{2h-1} = (-1)^{h+1}\frac{\sqrt{-1}}{(2\pi)^{2h-1}(2h-2)!}\cdot$$

$$\cdot\int_0^1\left\{\delta^{j_1\ldots j_{2h-1}}_{i_1\ldots i_{2h-1}}\alpha^{i_1}_{j_1}\wedge(\Theta_t)^{i_2}_{j_2}\wedge\ldots\wedge(\Theta_t)^{i_{2h-1}}_{j_{2h-1}}\right\}dt \quad .$$

Particularly, for $h=1$ we obtain

(4.4.44) $$\Delta(\theta_0,\theta_1)c_1 = \frac{\sqrt{-1}}{2\pi}\alpha^i_i = \frac{\sqrt{-1}}{2\pi}\left(\overset{1}{\nu}{}^i_h\, d\,\overset{1}{\gamma}{}^h_i - \overset{0}{\nu}{}^i_h\, d\,\overset{0}{\gamma}{}^h_i\right) =$$

$$= \frac{\sqrt{-1}}{4\pi}\, d\,\ln\frac{\det^2(\overset{1}{\gamma}{}^j_i)}{\det^2(\overset{0}{\gamma}{}^j_i)} \quad .$$

If we choose simply $\varepsilon_i = \overset{0}{\varepsilon}_i$, then $\overset{0}{\gamma} = I_n$ (unit matrix), and we see from (4.4.38) that $\overset{1}{\nu} = (\overset{1}{\gamma})^{-1}$ is precisely the matrix which defines the connecting mapping $A(L_0,L_1)$ of formula (4.3.3). Accordingly, formula (4.4.44) shows that the Maslov class of formula (4.3.5) is given by

(4.4.45) $$m(L_0,L_1) = \frac{1}{2}\mu_1(E,L_0,L_1) \quad ,$$

which justifies the name given to the classes μ_h. (An equivalent expression of $m(L_0,L_1)$ via connections has been given by P. Dazord [23].) Let us also notice that the above considerations (the choice $\varepsilon_i = \overset{0}{\varepsilon}_i$) yield one more fact, namely, if $\ell_1 : M \to \mathcal{L}(E)$ is the cross section of the Lagrange Grassmannian bundle of E which defines L_1 then

(4.4.46) $$\mu_1(E,L_0,L_1) = -\ell_1^*\,\bar{\mu}^1(E,L_0) \quad ,$$

where $\bar{\mu}^1(E,L_0)$ is the class (4.4.30). Indeed, it suffices to compare (4.4.44) and (4.4.29).

Now, we proceed with the basic properties of the classes μ_h.

Theorem 4.4.25. *The Maslov classes* $\mu_h(E,L_0,L_1)$ *do not depend on the choice of the* L_a*-orthogonal unitary connections* θ_a $(a = 0,1)$, *and they* *do not depend on the choice of the complex structure* J. *Furthermore,* $\mu_h(E,L_0,L_1)$ *are rigid with respect to homotopic deformations of* L_0 *and* L_1 *via Lagrangian subbundles of* E, *and* $\mu_h(E,L_0,L_1) = 0$ *if* L_0 *and* L_1 *are homotopic via Lagrangian subbundles.*

Proof. In the proof of Proposition 4.4.17, we saw that any two L-orthogonal unitary connections are c_{2h-1}-homotopic, L being a fixed Lagrangian subbundle of E. Hence, connections (θ_0,θ_1), (θ_0',θ_1') as needed for μ_h are c_{2h-1}-homotopic, respectively, and the classes μ_h which they define are equal by Theorem 4.2.16.

Two positive compatible complex structures J_0,J_1 are related by a family J_s of such structures, and this gives rise to the bundle $\mathcal{U}(J_0,J_1)$ of (4.4.12). It is easy to understand that we can define two O(n)-reductions of $\mathcal{U}(J_0,J_1)$ by

(4.4.47) $$\mathcal{W}_a = \bigcup_{x\in M,\, s\in[0,1]} \mathcal{U}(E,J_s;L_a)_x \qquad (a = 0,1).$$

Accordingly, connections χ_a on $\mathcal{W}_a$ extend to connections χ_a in $\mathcal{U}(J_0,J_1)$, and define forms $\Delta(\chi_0,\chi_1)c_{2h-1}$ which induce representative forms of $\mu_h(E,L_0,L_1)$ with respect to J_0 for s=0, and with respect to J_1 for s=1. But, these induced representative forms are cohomologous in view of the Stokes' formula (4.1.29) for $\alpha = \Delta(\chi_0,\chi_1)c_{2h-1}$ on $M\times I$. Hence, $\mu_h(E,L_0,L_1)$ do not depend on the choice of J. It is for this reason that we do not introduce J in the notation of the Maslov classes.

The rigidity of μ_h by homotopic deformations of L_a has a completely similar proof as Proposition 4.4.18, and we leave it to the reader. (One shall consider two subbundles (4.4.26) of $\mathcal{U}(E,J)\times I$ associated to $L_a(s)$, a=0,1, and apply Theorem 4.2.16 for simple secondary characteristic classes.) Now, the last assertion of our theorem follows by deforming (L_0,L_1) to (L_1,L_1), and by using Corollary 4.2.17. Q.e.d.

At this point, we can prove that the Maslov classes are obstructions to the transversality of L_0 and L_1, which was the motivation of the study of these classes. We have

Theorem 4.4.26. *If the Lagrangian subbundles* L_0, L_1 *of* E *are everywhere transversal then* $\mu_h(E, L_0, L_1) = 0$. *The same is true if* L_0, L_1 *can be deformed via Lagrangian subbundles to transversal bundles* L_0', L_1'.

Proof. If L_0, L_1 are transversal subbundles, we may choose J such that $JL_0 = L_1$. Accordingly, we may use equal connections $\theta_0 = \theta_1$ in the computation of $\mu_h(E, L_0, L_1)$. (Take $\overset{1}{e}_i = J\overset{0}{e}_i$ in (4.4.36), and $\underset{1}{\omega}{}^j_i = \underset{0}{\omega}{}^j_i$ in (4.4.37).) Hence $\mu_h(E, L_0, L_1) = 0$. The second assertion is clear from Theorem 4.4.25. Q.e.d.

Other interesting properties are given by

Proposition 4.4.27. *Let* L_α $(\alpha = 0,1,2)$ *be Lagrangian subbundles of the symplectic vector bundle* E. *Then we have*

$$\mu_h(E, L_0, L_1) = -\mu_h(E, L_1, L_0) \quad , \tag{4.4.48}$$

$$\mu_h(E, L_0, L_1) + \mu_h(E, L_1, L_2) + \mu_h(E, L_2, L_0) = 0 \quad . \tag{4.4.49}$$

Proof. The first equality is straightforward from (4.2.26). Then, the second relation follows by taking the L_α-orthogonal unitary connections θ_α $(\alpha = 0,1,2)$, and by using for them (4.2.13) with r=2. We get

$$d\big(\Delta(\theta_0, \theta_1, \theta_2)c_{2h-1}\big) = \Delta(\theta_1, \theta_2)c_{2h-1} -$$

$$- \Delta(\theta_0, \theta_2)c_{2h-1} + \Delta(\theta_0, \theta_1)c_{2h-1} ,$$

and (4.4.49) follows by going over to cohomology classes. Q.e.d.

Corollary 4.4.28. *If (up to homotopy via Lagrangian subbundles)* L_0 *and* L_1 *admit a common transversal Lagrangian subbundle* L_2, *then* $\mu_h(E, L_0, L_1) = 0$.

Proof. If L exists, the last two terms of (4.4.49) are zero. Q.e.d.

Hence, the Maslov classes are also obstructions to the existence of a global common transversal Lagrangian subbundle of two given Lagrangian subbundles. *This is another interesting geometric role of the Maslov classes.* (See [87], p. 26, for the significance of the existence of a common transversal Lagrangian subbundle.)

Proposition 4.4.29. *The Maslov classes, and the Chern-Simons-Maslov classes of two Lagrangian subbundles* L_0, L_1 *of* E *are related by*

(4.4.50) $$\mu^h(E,J;L_1) - \mu^h(E,J;L_0) = \pi^*\mu_h(E,L_0,L_1) \quad ,$$

where $\pi\colon \mathcal{U}(E,J) \to M$ *is the bundle projection.*

Proof. This is a particular case of Proposition 4.2.14. Q.e.d.

Corollary 4.4.30. *If* E *is a trivial symplectic bundle, then* $2\mu_h(E,L_0,L_1)$ *are real images of integral cohomology classes.*

Proof. Since E is trivial so is $\mathcal{U}(E,J)$, and we have a global cross section $s\colon M \to \mathcal{U}(E,J)$. Then, (4.4.50) implies

(4.4.51) $$\mu_h(E,L_0,L_1) = s^*\mu^h(E,J;L_1) - s^*\mu^h(E,J;L_0) \quad ,$$

and the result follows from Proposition 4.4.23. Q.e.d.

Now, following Hörmander [41, 37], let us lift the bundle E to $L(E)$ via the projection $\pi'\colon L(E) \to M$ of diagram (3.1.5). We get $\pi'^{-1}(E)$ which is a symplectic vector bundle $\hat{\pi}\colon \hat{E} \to L(E)$. $\hat{E}$ has a canonical Lagrangian subbundle S whose fibre $S(\ell)$ at $\ell \in L(E)$ is the Lagrangian subspace $\ell \subset E_{\pi(\ell)}$. We call S the *tautological subbundle* of $\hat{E}$.

Simultaneously, the U(n)-principal bundle $\mathcal{U}(E,J)$ lifts to $\hat{\mathcal{U}}(E,J)$, and we have a commutative diagram

$$
(4.4.52)\qquad \begin{array}{ccc} \mathcal{U}(E,J) & \overset{\tau}{\leftarrow} & \hat{\mathcal{U}}(E,J) \\ \pi\downarrow & & \downarrow\hat{\pi} \\ M & \underset{\pi'}{\leftarrow} & L(E) \end{array}
$$

Furthermore, for any Lagrangian subbundle, say, L_0 of E we get a lift $\hat{L}_0 = \pi'^{-1}L_0$ which is Lagrangian subbundle of $\hat{E}$.

Definition 4.4.31. *The cohomology classes* $\mu^h(\hat{E},J;S)$ *are called* ***universal Chern-Simons-Maslov classes of*** E. *The cohomology classes* $\mu_h(\hat{E},\hat{L}_0,S)$ *are called* ***universal Maslov classes of the pair*** (E,L_0).

These classes are global invariants of E, and of (E,L_0), respectively, and the name is explained by

Proposition 4.4.32. *The Chern-Simons-Maslov classes of any Lagrangian subbundle* L_1 *of* E, *and the Maslov classes of* L_1 *with respect to* L_0 *are induced by the corresponding universal classes via* L_1 *seen as a cross-section of* $L(E)$.

Proof. Let $\ell_1 : M \to L(M)$ be the section defined by L_1. Since $\pi' \circ \ell_1 = id_M$, we have identifications

$$
(4.4.53)\qquad \begin{array}{l} \ell_1^{-1}(\hat{E}) = E,\ \ell_1^{-1}\,\hat{\mathcal{U}}(E,J) = \mathcal{U}(E,J)\ , \\ \ell_1^{-1}(\hat{L}_0) = L_0\ ,\ \ell_1^{-1}(\hat{L}_1) = L_1\ ,\ \ell_1^{-1}(S) = L_1\ , \end{array}
$$

and we have corresponding commutative diagrams like

$$
(4.4.54)\qquad \begin{array}{ccc} \mathcal{U}(E,J) & \overset{\hat{\ell}_1}{\rightarrow} & \hat{\mathcal{U}}(E,J) \\ \pi\downarrow & & \downarrow\hat{\pi} \\ M & \underset{\ell_1}{\rightarrow} & L(E) \end{array}
$$

Now, from diagram (4.4.54), and Theorem 4.4.25, we get

(4.4.55) $$\mu^h(E,J;L_1) = \hat{\ell}_1^*\mu^h(\hat{E},J;S) ,$$

which is the first assertion of Proposition 4.4.32.

Furthermore, from (4.4.49), we get

(4.4.56) $$\mu_h(\hat{E},\hat{L}_0,\hat{L}_1) = \mu_h(\hat{E},S,\hat{L}_1) + \mu_h(\hat{E},\hat{L}_0,S) .$$

Here, we shall apply ℓ_1^*, and use again the diagram (4.4.54). If, we also notice that

$$\ell_1^*\mu_h(\hat{E},S,\hat{L}_1) = \mu_h(E,L_1,L_1) = 0 ,$$

then the result will be

(4.4.57) $$\mu_h(E,L_0,L_1) = \ell_1^*\mu_h(\hat{E},\hat{L}_0,S) .$$

Q.e.d.

Remark 4.4.33. *A comparison of formulas (4.4.29), and (4.4.43) for* $(\hat{L}_0,S)$ *shows that we have*

(4.4.58) $$\bar{\mu}^1(E,L_0) = -\mu_1(\hat{E},\hat{L}_0,S) ,$$

where $\bar{\mu}^1$ *is the class defined in (4.4.30).*

Remark 4.4.34. *There are also other interesting relations between the universal classes. For instance, from Proposition 4.4.29, we deduce*

(4.4.59) $$\mu^h(\hat{E},J;\hat{L}_0) - \mu^h(\hat{E},J;S) = \hat{\pi}^*\mu_h(\hat{E},S,\hat{L}_0) ,$$

and, in view of diagram (4.4.52) this means

(4.4.60) $$\tau^*\mu^h(E,J;L_0) = \mu^h(\hat{E},J;S) - \hat{\pi}^*\mu_h(\hat{E},\hat{L}_0,S) ,$$

etc.

We shall obtain another result about the Maslov classes by using a differentiable triangulation K of the base manifold M of the bundles in consideration, and the idea of Lehmann's definition of residues (Section 4.2). Then $\mu_h(E,L_0,L_1)$ is defined simplicially by the cocycle determined by $\int_\gamma \Delta(\theta_0,\theta_1)c_{2h-1}$ where γ runs over the set of (4h-3)-simplices of K. Every such simplex γ is a contractible space, hence we have cross sections of $U(E,J;L_a)/\gamma$ (a=0,1), and corresponding trivial connections (i.e., connections which make these cross sections parallel) that extend to connections φ_a on $U(E,J)/\gamma$ (a=0,1). From Proposition 4.2.3, it follows

$$(4.4.61)\qquad d\Delta(\varphi_0,\theta_0,\theta_1)c_{2h-1} = \Delta(\theta_0,\theta_1)c_{2h-1} - \Delta(\varphi_0,\theta_1)c_{2h-1} + \Delta(\varphi_0,\theta_0)c_{2h-1} .$$

If we integrate this equality along γ, and use Stokes' formula we get

$$(4.4.62)\qquad \int_\gamma \Delta(\theta_0,\theta_1)c_{2h-1} = \int_{\partial\gamma} \Delta(\varphi_0,\theta_0,\theta_1)c_{2h-1} + \int_\gamma \Delta(\varphi_0,\theta_1)c_{2h-1} - \int_\gamma \Delta(\varphi_0,\theta_0)c_{2h-1} .$$

Since φ_0 and θ_0 are both L_0-orthogonal connections for the same metric, the last term of (4.4.62) vanishes, and we get

$$(4.4.63)\qquad \int_\gamma \Delta(\theta_0,\theta_1)c_{2h-1} = \int_{\partial\gamma} \Delta(\varphi_0,\theta_0,\theta_1)c_{2h-1} + \int_\gamma \Delta(\varphi_0,\theta_1)c_{2h-1} .$$

In (4.4.63), we can compute the last integral by using similarly the triple of connections $(\varphi_0,\varphi_1,\theta_1)$. We obtain

$$(4.4.64)\qquad \int_\gamma \Delta(\varphi_0,\theta_1) = -\int_{\partial\gamma} \Delta(\varphi_0,\varphi_1,\theta_1)c_{2h-1} + \int_\gamma \Delta(\varphi_0,\varphi_1)c_{2h-1} + \int_\gamma \Delta(\varphi_1,\theta_1)c_{2h-1} .$$

Since φ_1, θ_1 are L_1-orthogonal connections, the last integral vanishes. Furthermore, $\Delta(\varphi_0, \varphi_1)c_{2h-1} = d\alpha$ for some form α since it is a closed form on a contractible set. By combining all these results we get

Proposition 4.4.35. *For every* (4h-3)*-dimensional simplex* γ *of* K, *it is possible to choose trivial* L_a*-orthogonal unitary connections* φ_a *on* $U(E,J)/\gamma$ (a=0,1), *and a form* α *such that* $\Delta(\varphi_0, \varphi_1)c_{2h-1} = d\alpha$, *and*

$$\int_\gamma \Delta(\theta_0, \theta_1)c_{2h-1} = \int_{\partial\gamma} \left\{\Delta(\varphi_0, \theta_0, \theta_1)c_{2h-1} - \Delta(\varphi_0, \varphi_1, \theta_1)c_{2h-1} + \alpha\right\}. \tag{4.4.65}$$

Remark 4.4.36. *The interest of formula (4.4.65) lies in the fact that it reduces the dimension of the integrals. For instance, if* h=1, *this formula yields a computation of the Maslov index of a closed curve which consists in adding up the "jumps" of some functions along the 1-simplices that compose the curve. We may also notice that the connections do not enter symmetrically in (4.4.65), but if we use (4.4.65), and if we use again Proposition 4.2.3, we get*

$$\int_\gamma \Delta(\theta_0, \theta_1)c_{2h-1} + \int_\gamma \Delta(\theta_1, \theta_0)c_{2h-1} =$$
$$= - \int_{\partial\gamma} d\Delta(\varphi_0, \varphi_1, \theta_0, \theta_1)c_{2h-1} = 0$$

as it should be.

In this context, it is perhaps the place to indicate also a relationship between Lehmann's residues and the transversality of Lagrangian subbundles, in spite of the fact that it is not via the Maslov classes. Therefore, let again L_0, L_1 be Lagrangian subbundles of the symplectic vector bundle $E \to M$, and let us assume that L_0, L_1 are transversal except for a closed subset $S \subset M$, called the *nontransversality set*. We shall also assume that there is a differentiable triangulation K of M such that $M \setminus S$ retracts onto a subcomplex $K' \subset K$, and such that for some integer $k > 0$ one has

(4.4.66) $$\text{skeleton}^{2k-1} K \cap S = \emptyset \, .$$

Now, let θ_0 be an arbitrary linear connection on L_0. Because of transversality, $L_1 \approx L_0^*$ along $M \setminus S$ (Section 3.1), and we get on L_1 an induced linear connection θ_0^* with the singular set S. Hence, there are well defined residues $\mathrm{Res}_\gamma(\theta_0^*, c_k) = \mathrm{Res}_\gamma(L_0, L_1, c_k)$, where γ is a 2k-dimensional simplex of K, and c_k is the Chern polynomial. By Proposition 4.2.31, in order to compute these residues, we have to take a linear connection without singularities θ_1 of L_1, and then we have

(4.4.67) $$\mathrm{Res}_\gamma(L_0, L_1, c_k) = \int_\gamma \Delta(\theta_1) c_k - \int_{\partial\gamma} \Delta(\theta_0^*, \theta_1) c_k \quad .$$

Furthermore, by Theorem 4.2.33, if c is a 2k-dimensional cycle of K, we get

(4.4.68) $$\mathrm{Res}_c (L_0, L_1, c_k) = \langle c_k(L_1), c \rangle = \langle c_k(E), c \rangle \quad ,$$

where $c_k(L_1)$ are the Chern classes of $L_1 \otimes \mathbb{C} \approx (E,J)$, which shows that the sum of residues along a cycle does not depend on the choice of θ_0.

For $k = 2h-1$, and if we use as θ_0, θ_1 connections that extend $O(n)$-connections on L_0, L_1, respectively, we are in the situation of Proposition 4.2.34, and we have cohomology classes

(4.4.69) $$[\mathrm{Res}(L_0, L_1, c_{2h-1})] = -\partial[\Delta(\theta_0^*, \theta_1) c_{2h-1}] \quad ,$$

where ∂ is the connecting homomorphism of the exact cohomology sequence of the pair (K, K'). It follows, just like for the Maslov classes (Theorem 4.4.25), that the classes (4.4.69) do not depend on the choice of the orthogonal connections θ_0, θ_1. It is clear that if $S = \emptyset$ we may take $\theta_1 = \theta_0^*$, and the classes (4.4.69) vanish. Therefore, these classes are again transversality obstructions.

Now, let us come back to the Maslov classes. Following the theory of Section 4.2, we may inquire about general secondary characteristic classes as introduced by Defintion 4.2.22. In a natural manner, such classes are associated to the ideals of $I(U(n))$ given by

(4.4.70) $J_0 = J_1 = \{c_1, c_3, \ldots\}$ = the ideal spaned by the Chern polynomials of odd degree.

Then, following Theorem 4.2.26, we have to consider the differential graded algebra

(4.4.71) $$WL_n \overset{def}{=} \mathbb{R}[c_2, c_4, \ldots] \otimes \mathbb{R}[c'_2, c'_4, \ldots] \otimes \wedge(u_1, u_2, \ldots, u_n) ,$$

where the generators c, c', u are different "copies" of the Chern polynomials, where the grades are defined by (4.2.52), and the differential d is given by $dc_{2h} = 0$, $dc'_{2h} = 0$, $du_{2h-1} = 0$, $du_{2h} = c'_{2h} - c_{2h}$. The secondary Chern-Weil homomorphism

(4.4.72) $$\rho(\theta_0, \theta_1): WL_n \to \wedge M$$

will be defined by (4.2.55) for L_a-orthogonal unitary connections θ_a $(a=0,1)$ on $\mathcal{U}(E,J)$ (the notation is, of course, the usual one in this Section), and it induces a cohomological homomorphism leading to the secondary characteristic classes of Definition 4.2.22.

In order to understand these classes, we shall proceed like in Theorem 6.1 of Lehmann's paper [49]. Namely, let us consider the subalgebra of WL_n given by

(4.4.73) $$\widehat{WL}_n = \mathbb{R}[c_2, c_4, \ldots] \otimes \mathbb{R}[c'_2, c'_4, \ldots] \otimes \wedge(u_1, u_3, \ldots) ,$$

which has the trivial differential $d=0$. Then, let U be the subalgebra of $\mathbb{R}[c_2, c_4, \ldots] \otimes \mathbb{R}[c'_2, c'_4, \ldots]$ generated by $\{du_2, du_4, \ldots\}$, and put

(4.4.74) $$V = U \otimes \wedge(u_2, u_4, \ldots)$$

which is again a graded differential subalgebra of WL_n. Now, we see that

$$WL_n = V \otimes_{\mathbb{R}} \hat{WL}_n \quad ,$$

and, by usual homological algebra (Künneth formula) we have

$$(4.4.75) \qquad H^j(WL_n) = \sum_{p+q=j} H^p(V) \otimes H^q(\hat{WL}_n) \quad .$$

The algebra V can be seen as a Koszul complex with cohomology spaces

$$(4.4.76) \qquad H^0(V) = \mathbb{R}\,,\ H^p(V) = 0 \quad \text{for} \quad p > 0\ .$$

The proof is by algebraic considerations which we prefer not to develop here, and which can be found in [49]. On the other hand, since $d=0$ on $\hat{WL}_n$, we have

$$(4.4.77) \qquad H^*(\hat{WL}_n) = \hat{WL}_n \quad .$$

Hence

$$(4.4.78) \qquad H^*(WL_n) = \hat{WL}_n \quad ,$$

and we proved

Theorem 4.4.37. *The general secondary characteristic classes associated to a pair* L_0, L_1 *of Lagrangian subbundles of a symplectic vector bundle* E *are real linear combinations of products of Pontrjagin classes of* L_0, *Pontrjagin classes of* L_1, *and Maslov classes* $\mu_h(E, L_0, L_1)$.

Accordingly all these classes depend only on the triple (E, L_0, L_1), and they vanish if L_0, L_1 are transversal subbundles. But, in fact, Theorem 4.4.37 proves that, in this case, the only interesting secondary characteristic classes are the Maslov classes.

4.5. COMPUTATIONS IN COTANGENT BUNDLES. The initial reason for the study of Maslov classes in this work was their application to Lagrangian submanifolds of cotangent bundles. The corresponding configuration has been described in Section 1.1. Namely, we consider the cotangent bundle T^*M of a differentiable manifold M^m, endowed with the canonical symplectic structure Ω defined by (1.1.5), and a Lagrangian submanifold N of T^*M, or a Lagrangian immersion $\iota : N \to T^*M$. Then, we define

$$E = \iota^{-1}T(T^*M), \quad L_0 = \iota^{-1}V, \quad L_1 = TN, \tag{4.5.1}$$

where V is the vertical foliation of T^*M, i.e., the tangent bundle of the fibers of T^*M. We are interested in the transversality of L_0, L_1 hence, by the results of Section 4.4, we shall be interested in the Maslov classes $\mu_h(E, L_0, L_1)$.

Definition 4.5.1. *The cohomology classes*

$$\mu_h(N) \overset{\text{def}}{=} \mu_h(E, L_0, L_1) \in H^{4h-3}(N, \mathbb{R}) \tag{4.5.2}$$

of the bundles defined in (4.5.1) are called ***the Maslov classes of*** N ***in*** T^*M.

Remark 4.5.2. *Similar to Definition 4.5.1, for any symplectic manifold* (V, ω) *endowed with a Lagrangian foliation* V, *if* $\iota : N \to V$ *is a Lagrangian immersion, the bundles (4.5.1) exist for* T^*M *replaced by* V, *and (4.5.2) define* ***Maslov classes of*** N ***in*** V, ***with respect to*** V. *On the other hand, let us remark that, as a consequence of formula (4.4.45),* $\mu_1(N) = (1/2)m(N)$, *where* $m(N)$ *is the usual Maslov class of* N *([37], etc.)*

In this Section, we develop a differential geometric machinery for the computation of the classes $\mu_h(N)$. Then, this machinery will be refined such as to give us also the Maslov classes of Legendrian submanifolds of cotangent unit spheres bundles.

The differential geometry of a cotangent bundle has been studied by many authors and the interested reader may find a systematic exposition of this subject in [90]. For the computation of Maslov classes (4.5.2) we need a compatible complex structure J on TT^*M, and the simplest way to obtain it is by using a fixed riemannian metric γ on M. Then we have [79]

Proposition 4.5.3. *The symplectic manifold* (T^*M,Ω), *where* M *is a riemannian manifold with metric* γ, *has a positive compatible almost complex structure* J *induced by* γ. *Hence, if* g *is the corresponding hermitian metric* (T^*M,J,g) *is an almost Kähler manifold. This manifold is Kähler iff* γ *is a flat metric.*

Proof. Let Γ be the Levi-Civita connection of γ. Then at each point $(x_0,\xi_0) \in T^*M$, we have a *horizontal space* H_0 which is defined as the space that contains the tangent vectors to the paths of T^*M obtained by parallel translations of ξ_0 in M with respect to Γ. By varying (x_0,ξ_0), we get thereby a *horizontal distribution* H on T^*M such that

$$(4.5.3) \qquad T(T^*M) = H \oplus V \ .$$

We shall see immediately that H is a Lagrangian subbundle, and γ defines a metric on H as well. Since V is also a Lagrangian subbundle, we have then a compatible complex structure J as defined in the first part of the proof of Theorem 2.3.3. This is the structure required in Proposition 4.5.3.

Now, for the details, let (x^i) $(i = 1,\dots,m)$ be local coordinates on M, and (x^i, ξ_i) be corresponding canonical cotangent coordinates on T*M, such that Ω is given by (1.1.5). For γ, we have $\gamma = \gamma_{ij} dx^i dx^j$, and the connection forms of Γ

$$\omega_i^j = \Gamma_{i\,k}^j dx^k \quad , \tag{4.5.4}$$

where $\Gamma^{\cdot}_{\cdot\cdot}$ are the Christoffel symbols of γ [31, 45]. Then, from the usual equations of a parallel translation [31, 45], we get for H the local equations

$$\theta_i \overset{def}{=} d\xi_i - \Gamma_{i\,h}^k \xi_k dx^h = 0 \quad , \tag{4.5.5}$$

which yields dim $H = n$. Since

$$\Gamma_{i\,h}^k = \Gamma_{hi}^k \quad , \tag{4.5.6}$$

we get from (1.1.5)

$$\Omega = dx^i \wedge d\xi_i = dx^i \wedge \theta_i \quad , \tag{4.5.7}$$

and this implies that H is a Lagrangian subbundle of TT*M. Of course, the same is true for V which is defined by $dx^i = 0$, and (4.5.3) becomes clear as well.

It is important to notice that we obtained above local bases in T*T*M given by (dx^i, θ_i), and, if $\tilde{x}^i = \tilde{x}^i(x^j)$ is a coordinate transformation on M, then we have

$$d\tilde{x}^i = \frac{\partial \tilde{x}^i}{\partial x^j} dx^j \ , \quad \tilde{\theta}_i = \frac{\partial x^h}{\partial \tilde{x}^i} \theta_h \quad . \tag{4.5.8}$$

The corresponding dual bases of TT*M are

$$X_i = \frac{\partial}{\partial x^i} + \Gamma_{i\,h}^k \xi_k \frac{\partial}{\partial \xi_h} \ , \quad \frac{\partial}{\partial \xi_i} \quad , \tag{4.5.9}$$

where the two groups of vectors span H and V respectively (i.e., these bases are *adapted to the almost product structure (4.5.3)* of T^*M), and we have transition relations induced by (4.5.8)

$$\tilde{X}_i = \frac{\partial x^h}{\partial \tilde{x}^i} X_h \quad , \quad \frac{\partial}{\partial \tilde{\xi}_i} = \frac{\partial \tilde{x}^i}{\partial x^h} \frac{\partial}{\partial \xi_h} \quad . \tag{4.5.10}$$

Now, we obtain a riemannian metric γ on the bundle H by

$$\gamma_H(v,w) = \gamma(\pi_* v, \pi_* w) \quad , \tag{4.5.11}$$

where $v,w \in H$, π: $T^*M \to M$. It is easy to see that γ_H is well defined, and

$$\gamma_H(X_i, X_j) = \gamma_{ij} \quad . \tag{4.5.12}$$

Accordingly, we may proceed like in the first part of the proof of Theorem 2.3.3. Namely, let us consider in H the local bases

$$Y^i = \gamma^{ih} X_h \qquad (\gamma^{ih}\gamma_{hk} = \delta^i_k) \quad , \tag{4.5.13}$$

whose transition relations induced by (4.5.10) are

$$\tilde{Y}^i = \frac{\partial \tilde{x}^i}{\partial x^j} Y^j \ . \tag{4.5.14}$$

Then, since H^* has the basis dx^i, it follows from (4.5.8) and (4.5.14) that the isomorphism c: $H \to H^*$ is defined by $c(Y^i) = dx^i$. Now, it follows from (4.5.10), (4.5.14) that the operator J defined by

$$JY^h = \frac{\partial}{\partial \xi_h} \ , \ J \frac{\partial}{\partial \xi_h} = - Y^h \tag{4.5.15}$$

has an invariant meaning namely, $J|_H = -b^{-1} \circ c$, where b is Ω-duality. It is also clear from (4.5.15) that J is a complex structure on TT^*M i.e., J is an almost complex structure on T^*M [45].

It is easy to check the compatibility relation

$$\Omega(Jv, Jw) = \Omega(v,w) \quad , \tag{4.5.16}$$

and the corresponding hermitian metric of T^*M is

$$g(v,w) = \Omega(v,Jw) \tag{4.5.17}$$

which implies

$$g(Y^i,Y^j) = g\left(\frac{\partial}{\partial \xi_i}, \frac{\partial}{\partial \xi_j}\right) = \gamma^{ij}, \quad g\left(Y^i, \frac{\partial}{\partial \xi_j}\right) = 0 ,$$

(4.5.18)

$$g(X_i,X_j) = \gamma_{ij} \quad .$$

Hence g is a positive definite metric, whose restriction to H equals γ_H, and H, V are g-orthogonal.

This proves the first part of Proposition 4.5.3. J of (4.5.15) is the structure induced by γ, and T^*M becomes an almost Kähler manifold.

Furthermore, M is Kähler iff J is an integrable structure i.e., it defines a complex structure on T^*M, and it is known that this happens iff its Nijenhuis tensor vanishes [45] i.e.,

$$N(v,w) = [Jv,Jw] - [v,w] - J[v,Jw] - J[Jv,w] = 0 . \tag{4.5.19}$$

It is easy to see that (4.5.19) is equivalent to the local condition

$$[Y^k,Y^h] = (\gamma^{hi}\Gamma^k_{i\ell} - \gamma^{ki}\Gamma^h_{i\ell})Y^\ell \tag{4.5.20}$$

i.e., $N=0$ iff the horizontal distribution is integrable. But from the equations (4.5.5) of H we get by a straightforward computation

$$d\theta_i = -\frac{1}{2} R^k_{i\ell h}\xi_k dx^\ell \wedge dx^h \quad (\text{mod } \theta_i = 0) , \tag{4.5.21}$$

where R is the curvature tensor of the metric γ given by [31, 45]

$$R^k_{i\ell h} = \frac{\partial \Gamma^k_{ih}}{\partial x^\ell} - \frac{\partial \Gamma^k_{i\ell}}{\partial x^h} + \Gamma^p_{ih}\Gamma^k_{p\ell} - \Gamma^p_{i\ell}\Gamma^k_{ph} \quad . \tag{4.5.22}$$

Hence the classical Frobenius theorem [59] shows that H is integrable iff $R=0$ i.e., g is flat. Q.e.d.

For the computation of the Maslov classes we also need adequate connections, and we shall prove now

Proposition 4.5.4. *There is a well defined complex metric connection* $\hat{\Gamma}$ *on* (T^*M,J,g) *such that the* $\hat{\Gamma}$*-parallel translations preserve the vertical and horizontal distributions* V,H *and the torsion of* $\hat{\Gamma}$ *has no horizontal component.* $\hat{\Gamma}$ *and* γ *are simultaneously flat, and then* $\hat{\Gamma}$ *is precisely the Levi-Civita connection* G *of the metric* g.

Proof. Since we want V and H to be parallel distributions, if $\hat{\Gamma}$ exists, it must have local equations of the form

$$\nabla Y^i = \alpha^i_j Y^j,\ \nabla \frac{\partial}{\partial \xi_i} = \beta_{ij} \frac{\partial}{\partial \xi_j}, \tag{4.5.23}$$

where α,β are the connection forms of $\hat{\Gamma}$.

Moreover, $\hat{\Gamma}$ is complex iff ∇ commutes with J which, in view of (4.5.15), is equivalent to

$$\beta_{ij} = \alpha^i_j \quad . \tag{4.5.24}$$

In (4.5.23), we used the local bases $\left(Y^i, \frac{\partial}{\partial \xi_j}\right)$ of TT^*M. The corresponding dual cobases are

$$\eta_i = \gamma_{ih} dx^h\ ,\ \theta_i\ , \tag{4.5.25}$$

and the connection forms can be expressed as

$$\alpha^i_j = A^{ih}_j \eta_h + \tilde{A}^{ik}_j \theta_k \quad . \tag{4.5.26}$$

It is known from the general theory of linear connections [31, 45] that the *torsion* of $\hat{\Gamma}$ has an expression

(4.5.27) $$T = E_i Y^i + \Theta_i \frac{\partial}{\partial \xi_i} ,$$

where the *torsion 2-forms* E_i, Θ_i are given by

(4.5.28) $$E_i = d\eta_i - \eta_h \wedge \alpha^h_i \quad , \quad \Theta_i = d\theta_i - \theta_h \wedge \alpha^h_i \quad .$$

The hypothesis about the torsion of $\hat{\Gamma}$ means $E_i = 0$ and by computing with respect to the cobasis (4.5.25) this means

$$\tilde{A}^{ik}_j = 0; \quad A^{hi}_j - A^{ih}_j = \gamma^{hk}(Y^p \gamma_{ih}) -$$

$$- \gamma^{hp}(Y^k \gamma_{ih}) = \gamma^{kh}\gamma^{iq}(X_q \gamma_{jk} - X_h \gamma_{jq}) .$$

Now, in order to give a short presentation, it may be proven by a straightforward computation that these conditions are fulfilled by taking

(4.5.29) $$\alpha^i_j = - \omega^i_j ,$$

where ω^i_j are the connection forms (4.5.4) of the Levi-Civita connection Γ of γ. (Notice that this means $A^{ih}_j = -\Gamma^i_{jk}\gamma^{kh}$.) This means that we take $\hat{\Gamma}$ defined by

(4.5.30) $$\nabla Y^i = -\omega^i_j Y^j \ , \ \nabla \frac{\partial}{\partial \xi_i} = - \omega^i_j \frac{\partial}{\partial \xi_j} ,$$

which is an easy guess since we want $\hat{\Gamma}$ to be a metric connection.

And, indeed, it is easy to see that formulas (4.5.30) yield a well defined metric connection on T^*M. This proves the first part of Proposition 4.5.4.

For the choice (4.5.29), the second group of torsion forms (4.5.28) is now seen to be

(4.5.31) $$\Theta_i = - \frac{1}{2} R^{\ell}_{ikh}\xi_{\ell} dx^k \wedge dx^h .$$

Hence, we see that the metric γ is flat (i.e., R=0), iff the metric connection $\hat{\Gamma}$ has no torsion i.e., iff $\hat{\Gamma}$ is precisely the Levi-Civita connection of g.

Finally, it follows from (4.5.30) and (4.1.21) that the matrix of the curvature forms of $\hat{\Gamma}$ with respect to the bases $\left(Y^i, \frac{\partial}{\partial \xi_j}\right)$ is

$$(4.5.32) \qquad \hat{\Omega} = \begin{pmatrix} -\Omega^i_j & 0 \\ 0 & -\Omega^i_j \end{pmatrix}$$

where

$$(4.5.33) \qquad \Omega^j_i = d\omega^j_i - \omega^h_i \wedge \omega^j_h$$

are the curvature forms of the riemannian curvature of γ. Hence, obviously, $\hat{\Gamma}$ and γ are simultaneously flat. Q.e.d.

Definition 4.5.5. *The connection* $\hat{\Gamma}$ *will be called the* ***complex lift*** *of the Levi-Civita connection* Γ *of* γ *to* T^*M.

Remark 4.5.6. *The connection* $\hat{\Gamma}$ *is precisely the* ***second connection*** *of the riemannian manifold* (T^*M,g) *with respect to the foliation* V*, in the sense of* [80].

Now, let us refer to a Lagrangian immersion $\iota : N \to T^*M$, where T^*M is endowed with the almost Kähler structure (J,g). Let $\hat{\nabla}$ denote the covariant derivative of the connection $\hat{\Gamma}$ of Proposition 4.5.4. Then, for vector field X,Y in TN we have a decomposition

$$(4.5.34) \qquad \hat{\nabla}_X Y = \nabla'_X Y + \hat{\sigma}(X,Y)$$

where the first term is in TN, and the second term is in the g-normal space $(TN)^{\perp}$. Equation (4.5.34) is similar to the Gauss equation in riemannian submanifolds geometry [45], and we call it the $\hat{\Gamma}$-*Gauss equation* of N. Like in the usual Gauss equations ∇' is an *induced metric* *connection on* N with respect to the metric ι^*g, but it may have torsion, and $\hat{\sigma}$ is a $(TN)^{\perp}$-valued 2-covariant tensor on N, which is not necessarily symmetric, and which we call the $\hat{\Gamma}$-*second fundamental form of* N.

The $\hat{\Gamma}$-Gauss equation (4.5.34) can be also expressed by means of local frames. Namely, let (e_i) $(i = 1,\dots,n)$ be a local field of g-orthonormal tangent frames of N. Then, since J is Ω-compatible, (Je_i) $(i = 1,\dots,n)$ is a local orthonormal frame field g-normal to N, and the decomposition (4.5.34) becomes

$$(4.5.35) \qquad \hat{\nabla}e_i = \lambda_i^j e_j + b_i^j(Je_j) \quad ,$$

where (λ_i^j) are the local connection forms of the induced connection ∇', and (b_i^j) are the local forms of $\hat{\sigma}$. Let us also remark that, since $\hat{\nabla}$ commutes with J, (4.5.35) implies

$$(4.5.36) \qquad \hat{\nabla}(Je_i) = -b_i^j e_j + \lambda_i^j(Je_j) \quad ,$$

which can be seen as the $\hat{\Gamma}$-*Weingarten equations of* N [45].

Proposition 4.5.7. *The coefficients of equation (4.5.35) satisfy the conditions*

$$(4.5.37) \qquad \lambda_i^j + \lambda_j^i = 0 \quad , \quad b_i^j = b_j^i \quad .$$

Proof. The first relation (4.5.37) follows since ∇' is a metric connection, and (e_i) is an orthonormal frame field. Furthermore, since $\hat{\Gamma}$ is a metric connection we have

$$0 = d\big(g(e_i,Je_j)\big) = g\big(\hat{\nabla}e_i,Je_j\big) + g\big(e_i,\hat{\nabla}(Je_j)\big) \;,$$

and if we use here (4.5.35) and (4.5.36) we get $b_i^j = b_j^i$. Q.e.d.

We may continue to mimic the riemannian submanifolds theory by establishing $\hat{\Gamma}$-*Gauss-Codazzi integrability conditions* [45]. Namely, by

(4.5.35), (4.5.36) the local connection matrix of $\hat{\Gamma}$ is

$$(4.5.38)\qquad \begin{pmatrix} \lambda_i^j & b_i^j \\ -b_i^j & \lambda_i^j \end{pmatrix}$$

and the corresponding matrix of curvature forms can be written as

$$(4.5.39)\qquad \begin{pmatrix} A_i^j & B_i^j \\ C_i^j & D_i^j \end{pmatrix} .$$

(This is the same curvature as (4.5.32) but with respect to other local bases.) Now, by a straightforward use of formulas (4.1.21), (4.1.22), (4.5.38), (4.5.39), we obtain

Proposition 4.5.8. *The curvature forms (4.5.39) satisfy the following* $\hat{\Gamma}$*-Gauss-Codazzi conditions*

$$(4.5.40)\qquad A_i^j = D_i^j = \Lambda_i^j + b_i^h \wedge b_h^j , \quad B_i^j = -C_i^j = (\nabla' b)_i^j ,$$

where

$$(4.5.41)\qquad \Lambda_i^j = d\lambda_i^j - \lambda_i^h \wedge \lambda_h^j$$

is the inner curvature of N *i.e., the curvature of* ∇', *and* $\nabla' b$ *is the covariant curvature of the second fundamental form* b *on* N.

Finally, let us go over to Maslov classes. We have

Theorem 4.5.9. *The Maslov classes* $\mu_h(N)$ *are expressible in terms of the inner curvature and the* $\hat{\Gamma}$*-second fundamental form of* N, *and its covariant derivative. If* N *is such that* $\hat{\sigma} = 0$, *then* $\mu_h(N) = 0$.

Proof. The general expression of the Maslov classes is given by formula (4.4.43). In our case, the elements which compose (4.4.43) can be computed as follows. E, L_0 and L_1 are given by (4.5.1). The basis (e_i) of (4.5.35) can be used in order to construct the L_1-related unitary frames $(\overset{1}{\varepsilon}_i)$ of (4.4.36), and, moreover, we shall use as a common basis (ε_i) of (4.4.38) the basis $\varepsilon_i = \overset{1}{\varepsilon}_i$. As the L_1-orthogonal unitary connection θ_1 of (4.4.37), we may use the induced connection ∇', hence with the notation of (4.4.39) and (4.5.35) we will have

$$\overset{1}{\pi}{}_i^j = \lambda_i^j \quad . \tag{4.5.42}$$

Furthermore, we see from the defining properties of the connection $\hat{\Gamma}$, that we may use the complex form of this connection to play the role of the L_0-orthogonal unitary connection θ_0 of (4.4.37). With respect to the bases (ε_i), it will have the local equations

$$\hat{\nabla}\varepsilon_i = \overset{0}{\pi}{}_i^j \varepsilon_j \quad , \tag{4.5.43}$$

where $\overset{0}{\pi}{}_i^j$ are the forms needed in (4.4.39). Since

$$\varepsilon_i = \frac{1}{\sqrt{2}}(e_i - \sqrt{-1}\,Je_i) \quad , \tag{4.5.44}$$

by comparing (4.5.35), (4.5.36), and (4.5.43), we get

$$\overset{0}{\pi}{}_i^j = \lambda_i^j + \sqrt{-1}\,b_i^j \quad . \tag{4.5.45}$$

Accordingly, from (4.5.42) and (4.5.45), we get the forms α_i^j nedded in (4.4.43). These forms are

$$\alpha_i^j = \overset{1}{\pi}{}_i^j - \overset{0}{\pi}{}_i^j = -\sqrt{-1}\,b_i^j \tag{4.5.46}$$

This result already gives for $\mu_1(N)$ the representative form (4.4.44) as

$$\Delta(\theta_0,\theta_1)c_1 = \frac{1}{2\pi}\,b_i^i \quad , \tag{4.5.47}$$

which proves our theorem for h=1, and, moreover, in this case, the curvature plays no role.

Let us also notice that (4.5.46) and (4.4.43) entirely prove the second assertion of Theorem 4.5.9.

Furthermore, we need now the curvature forms (4.4.41). From (4.4.41), (4.5.42) and (4.5.45), we obtain by a straightforward computation that

$$(4.5.48) \qquad (\Theta_t)_i^j = \Lambda_i^j + (1-t)^2 b_i^h \wedge b_h^j + \sqrt{-1}(1-t)(\nabla' b)_i^j \quad ,$$

which completes the proof. Q.e.d.

Remark 4.5.10. *In view of the Gauss-Codazzi conditions (4.5.40), (4.5.48) is also equivalent to another interesting formula namely*

$$(4.5.49) \qquad (\Theta_t)_i^j = A_i^j + \sqrt{-1}\,(1-t)B_i^j + t(t-2)b_i^h \wedge b_h^j \quad .$$

Interesting formulas are obtained in particular cases as follows

Theorem 4.5.11. *If either i)* $\hat{\Gamma}$ *induces in* N *a flat connection with respect to which the* $\hat{\Gamma}$*-second fundamental form is parallel, or ii)* M *is a flat riemannian manifold then* $\mu_h(N) = [m_h]$ *where*

$$(4.5.50) \qquad m_h = \frac{1}{(2\pi)^{2h-1}(2h-2)!} \nu_h \delta^{j_1 \dots j_{2h-1}}_{i_1 \dots i_{2h-1}} b_{j_1}^{i_1} \wedge b_{j_2}^{k_2} \wedge b_{k_2}^{i_2} \wedge \dots \wedge b_{j_{2h-1}}^{k_{2h-1}} \wedge b_{k_{2h-1}}^{i_{2h-1}} \quad ,$$

and ν_h *are respectively equal to*

i) $$\nu_h = \frac{(-1)^{h+1}}{4h-3}$$

ii) $$\nu_h = \sum_{i=0}^{2h-2} (-1)^{h+i+1} \frac{2^i}{4h-i-3} \binom{2h-2}{i} .$$

Proof. In case i), we have to use (4.5.48) with $\Lambda_i^j = (\nabla' b)_i^j = 0$. In case ii) we have to use (4.5.49) with $A_i^j = B_i^j = 0$ which holds because the flatness of M implies the flatness of $\hat{\Gamma}$ (Proposition 4.5.4). If these results are introduced in (4.4.43), one gets precisely (4.5.50). Q.e.d.

The second case of Theorem 4.5.11 is more interesting since in this case (T^*M,J,g) is a flat Kähler manifold, and $\hat{\Gamma}$ is its Levi-Civita connection. The $\hat{\Gamma}$-second fundamental forms is simply the usual second fundamental form of the submanifold N, and formula (4.5.50) has a clearer geometric significance. The case $M = \mathbb{R}^n$ with the canonical flat metric γ is included here. In this case, (T^*N,J,g) is precisely $\mathbb{C}^n$ with the canonical flat Kähler metric, and we have an explicit formula for all the Maslov classes of the Lagrangian submanifolds of $\mathbb{C}^n$. A similar formula has been given by Morvan and Niglio [66]. (Let us notice that, here, $\mathbb{C}^n = \{(z^i = x^i + \sqrt{-1}\, y^i)\}$, and the vertical distribution is x^i = const., while the usual Maslov class is computed with respect to the planes y^i = const. (Section 4.3). Since x^i = const., y^i = const. are transversal planes it follows from (4.4.49) that we obtain the same results.) Particularly, we may reprove the following result of Morvan [62]

Proposition 4.5.12. *Let* N *be a Lagrangian submanifold of* T^*M*, where* M *is a flat riemannian manifold and let* H *be the mean curvature vector of* N*. Then*

$$(4.5.51) \qquad \mu_1 = -\left[\frac{n}{2\pi}\, i(H)\Omega\right] ,$$

and $\mu_1 = 0$ *if* N *is a minimal submanifold.*

Proof. The mean curvature vector is defined as $1/n$ times the trace of the second fundamental form of N, hence, with the notation of (4.5.35), we have

$$(4.5.52)\qquad H = \frac{1}{n}\,\mathrm{tr}\,\hat{\sigma} = \frac{1}{n}\sum_{i=1}^{n}\hat{\sigma}(e_i,e_i) = \frac{1}{n}\sum_{i,j=1}^{n} b_i^j(e_i)(Je_j)\ ,$$

and, in view of (4.5.37)

$$(4.5.53)\qquad H = \frac{1}{n}\sum_{i,j=1}^{n} b_j^i(e_i)(Je_j)\ .$$

Furthermore, (4.5.35) implies

$$(4.5.54)\qquad b_j^i(e_k) = g(\hat{\nabla}_{e_k} e_j\ ,\ Je_i)\ ,$$

and since in our case $\hat{\nabla}$ has no torsion, it is easy to get

$$(4.5.55)\qquad b_j^i(e_k) = b_k^i(e_j).$$

Using this result, (4.5.53) becomes

$$(4.5.56)\qquad H = \frac{1}{n}\sum_{i,j=1}^{n} b_i^i(e_j)Je_j\ ,$$

and a straightforward computation gives

$$(4.5.57)\qquad i(H)\Omega = -\frac{1}{n}\sum_{i=1}^{n} b_i^i\ .$$

A comparison with (4.5.47) yields the result. Q.e.d.

Corollary 4.5.13. *Let* (V^{2n},J,g) *be a Kähler manifold which admits a parallel Lagrangian distribution* L_0*, and let* N *be an immersed Lagrangian submanifold of* V*. Then, the Maslov classes* $\mu_h(TV/N,L_0/N,TN)$ *are represented by the same forms* m_h *given by (4.5.50,ii), and, particularly* μ_1 *is given by (4.5.51).*

Proof. This case is more general since, for instance, V may be a compact manifold. However, we shall see that V must be flat.

Since L_0 is parallel with respect to a torsionless connection it follows from the definition of the torsion tensor [45] that L_0 is a foliation, and so is its orthogonal distribution $L_0^{\perp}$. Accordingly, g is a *reducible metric* [45] i.e., every point of V has a neighbourhood endowed with real local coordinates (x^a, y^α) $(a,b,\ldots = 1,\ldots,n;$ $\alpha,\beta,\ldots = 1,\ldots,n)$ such that $dx^a = 0$ defines L_0, $dy^\alpha = 0$ defines $L_0^{\perp}$, and g has the local expression

$$g = g_{ab}(x)dx^a dx^b + \tilde{g}_{\alpha\beta}(y)dy^\alpha dy^\beta \, . \tag{4.5.58}$$

Moreover, since $L_0, L_0^{\perp}$ are Lagrangian distributions, the Kähler form of V must have the local expression

$$\Omega = \omega_{a\alpha}(x,y)dx^a \wedge dy^\alpha \, . \tag{4.5.59}$$

Now, the fact that V is a Kähler manifold is equivalent to $\nabla\Omega = 0$, where ∇ is the Levi-Civita connection of g [45], and, using the index notation of covariant derivatives [31], this means

$$\nabla_b \omega_{a\alpha} = 0, \; \nabla_\beta \omega_{a\alpha} = 0. \tag{4.5.60}$$

From (4.5.60), and in view of the classical commutation formulas for covariant derivatives [31], we get

$$\nabla_c \nabla_b \omega_{a\alpha} - \nabla_b \nabla_c \omega_{a\alpha} = 0 = R^{e}_{acb}\omega_{e\alpha} \, , \tag{4.5.61}$$

where R is the curvature tensor of ∇. Therefore $R^{e}_{acb} = 0$. Similarly, (4.5.60) also implies $R^{\varepsilon}_{\alpha\beta\gamma} = 0$, and because of (4.5.58), we see that V is a flat Kähler manifold.

Then, it is well known that (V,J,g) is locally equivalent to $\mathbb{C}^n$. Since the computation of the representative differential forms of the Maslov classes has a local character, and, since the result is true for $\mathbb{C}^n$ by Theorem 4.5.11, and Proposition 4.5.12, we are done. Q.e.d.

Let us also reprove the following generalization of Proposition 4.5.12, due to P. Dazord [23]

Proposition 4.5.14. *Let* (V^{2n},J,g) *be a simply connected Kähler manifold with zero Ricci curvature* [45], *and endowed with an arbitrary Lagrangian distribution* L_0. *Let* N *be a Lagrangian submanifold of* V *with the mean curvature vector* H. *Then*

$$(4.5.62)\qquad \mu_1(TV/N, L_0/N, TN) = -\left[\frac{n}{2\pi}\, i(H)\Omega\right] ,$$

where Ω *is the Kähler form of* V.

Proof. As in (4.4.44), we represent μ_1 of (4.5.62) by

$$(4.5.63)\qquad \Delta(\theta_0,\theta_1)c_1 = \frac{\sqrt{-1}}{2\pi}\,\alpha^i_i = -\frac{\sqrt{-1}}{2\pi}\, d \ln \det (\xi^j_i) ,$$

where α^i_i is given by (4.4.40), the connections θ_0,θ_1 are induced by the Levi-Civita connection ∇ of V, and (ξ^j_i) is the matrix of the connecting mapping of L_0, L_1 as defined in formula (4.3.2).

First, we can write here again ∇ under the form of the Gauss-Weingarten equations (4.5.35), (4.5.36) of N in V, where we put ∇ instead of $\hat{\nabla}$. Then, the complex connection forms of the complex structure of TV/N are precisely the forms (4.5.45), which we denote now as

$$(4.5.64)\qquad \omega^j_i = \lambda^j_i + \sqrt{-1}\, b^j_i ,$$

where λ and b satisfy (4.5.37). As a consequence, we see that the Ricci form of V along N is [45]

$$(4.5.65)\qquad S/N = \sqrt{-1}\, db^i_i = 0$$

where S=0 is by the hypothesis of the proposition.

Now, exactly the same computation holds if TN is replaced by L_0 i.e., by using in (4.5.35) and (4.5.36) L_0-related unitary frames

(e'_i, Je'_i) instead of (e_i, Je_i). Then we obtain the new components of ∇ under the form

$$\omega'^j_i = \lambda'^j_i + \sqrt{-1}\, b'^j_i \tag{4.5.66}$$

and $S=0$ now means

$$d\omega'^i_i = \sqrt{-1}\, db'^i_i = 0 \ . \tag{4.5.67}$$

The relation (4.5.67) holds over the whole manifold V, and since the latter is simply connected we must have

$$\omega'^i_i = d\alpha \tag{4.5.68}$$

where α is a function on V. ω'^i_i is a well defined global 1-form if we use always L_0-orthogonal unitary frames.

Furthermore, along N we have

$$\varepsilon_i = \xi^j_i \varepsilon'_j \ , \tag{4.5.69}$$

where (ξ^j_i) is the unitary matrix of formula (4.5.63), and accordingly (see (4.1.18))

$$\omega^j_i = \eta^j_h d\xi^h_i + \eta^j_h \xi^k_i \omega'^h_k \ , \tag{4.5.70}$$

where $(\eta) = (\xi)^{-1}$. From (4.5.70) we deduce

$$\omega^i_i = d \ln \det (\xi^j_i) + \omega'^i_i \ . \tag{4.5.71}$$

Finally, in view of (4.5.63), (4.5.64), and (4.5.68), the relation (4.5.71) becomes

$$\Delta(\theta_0, \theta_1) c_1 = \frac{1}{2\pi} b^i_i + \frac{\sqrt{-1}}{2} d\alpha \ . \tag{4.5.72}$$

Since (4.5.57) clearly holds in our case as well, we are done. Q.e.d.

Finally, it is interesting to show that the same computation which led to formula (4.5.50ii) can be used also in a different situation that provides us with an example of nonvanishing Maslov classes. The example itself is the one used by Morvan and Niglio [66]. Namely, we consider the trivial symplectic bundle with an already fixed compatible complex structure $E = L(\mathbb{C}^n) \times \mathbb{C}^n$ whose basis is the Lagrange Grassmannian of the complex space $\mathbb{C}^n = \{(z^h = x^h + \sqrt{-1}\, y^h)\}$ $(h = 1,\ldots,n)$. Then, let L_0 be the trivial Lagrangian subbundle of E defined by the horizontal plane $y^h = 0$ of $\mathbb{C}^n$, and let S be the tautological Lagrangian subbundle that associates to $L \in L(\mathbb{C}^n)$ the plane L of $\mathbb{C}^n$ (Section 4.4). We want to compute the Maslov classes $\mu_h(E,L_0,S)$, which, in fact, are the universal Maslov classes of $(\mathbb{C}^n, L_0) \to \{\text{point}\}$ in the sense of Definition 4.4.31.

In order to do this, we take (e_i) $(i = 1,\ldots,n)$ to be a local orthonormal basis of S, and we consider the trivial connection $\hat{\nabla}$ of E. Then, we have Gauss-Weingarten equations (4.5.35), (4.5.36) of S in E, and the consequences (4.5.37), (4.5.40) of these equations. Moreover, in (4.5.40), $A^j_i = B^j_i = 0$ since $\hat{\nabla}$ has no curvature. By (4.5.35), $\hat{\nabla}$ induces a connection ∇' with the local forms (λ^j_i) in L, which is a metric connection, and may be used to define θ_1 needed in the computation of the Maslov classes. From the same formula, we also obtain the second fundamental form (b^j_i) of S in E. Furthermore, the connection θ_0 needed in the computation of $\mu_h(E,L_0,S)$ may be defined by $\hat{\nabla}$ itself, and if we go over to the basis ε_i of (4.5.44), we refind (4.5.45), (4.5.46), and, therefore, we also refind (4.5.50ii) as claimed earlier.

But in the present case, (b^j_i) are forms on $L(\mathbb{C}^n)$ which can be interpreted as follows. Let

$$f_i, Jf_i,\ \varphi_i = f_i - \sqrt{-1}\, Jf_i \qquad (i = 1,\ldots,n)$$

be the canonical basis of $\mathbb{C}^n$. Then

$$\varepsilon_i = u^j_i \varphi_j \qquad (u^j_i = p^j_i + \sqrt{-1}\, q^j_i) \ ,$$

where $(u^j_i) \in U(n)$, and it represents $L \in \mathcal{L}(\mathbb{C}^n)$ in the known identification of $(\mathbb{C}^n)$ with $U(n)/O(n)$. Accordingly, we get

$$e_i = p^j_i f_j + q^j_i (Jf_j), \ Je_i = -q^j_i f_j + p^j_i (Jf_j) \ ,$$

$$f_i = \sum_{j=1}^{n} p^i_j e_j - \sum_{j=1}^{n} q^i_j (Je_j), \ Jf_i = \sum_{j=1}^{n} q^i_j e_j + \sum_{j=1}^{n} p^i_j (Je_j) \ ,$$

and since $\hat{\nabla} f_i = \hat{\nabla}(Jf_i) = 0$, we can compute the coefficients of the Gauss equations (4.5.35). Particularly, we get (using also the symmetry of b)

$$b^j_i = \sum_{h=1}^{n} p^h_j dq^h_i - \sum_{h=1}^{n} q^h_j dp^h_i = \mathrm{Im}\ (u^{-1} du)^j_i \quad ,$$

where $u = (u^j_i) \in U(n)$. Therefore, b is the imaginary part of the Maurer-Cartan form of the unitary group $U(n)$, and a simple comparison of (4.5.50) with the known formula of the generators of $H^*(U(n)/O(n), \mathbb{R})$ as written for instance in Section 2.6 of [66] (see also [34]), shows that $\mu_h(E, L_0, S)$ are proportional to the generators of $H^*(U(n)/O(n), \mathbb{R})$, and particularly $\mu_h(E, L_0, S) \neq 0$.

Now, we shall discuss similar computations for Legendrian submanifolds of unit spheres cotangent bundles. Partly, these computations were presented in [84].

Hence, let M^{n+1} be a riemannian manifold with metric γ, and S^*M be the manifold of the cotangent unit spheres of M as defined in (3.2.22) i.e., the submanifold defined in T^*M by

$$\text{(4.5.73)} \qquad \gamma(\xi, \xi) = \gamma^{\alpha\beta} \xi_\alpha \xi_\beta = 1 \quad (\alpha, \beta, \ldots, 1, \ldots, n+1) \ .$$

This is a contact manifold with contact form (3.2.23) i.e.

(4.5.74) $\eta = \xi_i dx^i$ (modulo $\gamma(\xi,\xi) = 1$) ,

with the contact distribution C defined by $\eta = 0$, with the characteristic vector field $E = A/S^*M$ computed in (3.2.25), and with the symplectic structure

(4.5.75) $d\eta/C = -(\text{pullback } \Omega \text{ to } S^*M)/C$.

(Notice the change of sign in (4.5.75). Let us also remember from Section 3.3 that the fibers of S^*M, define the Legendrian vertical foliation C i.e., TC is a Lagrangian subbundle of C.

Now, let Λ^n be a Legendrian submanifold of S^*M (Definition 3.3.13). Then, we may consider the symplectic vector bundle E, and the Lagrangian subbundles L_0, L_1 defined by

(4.5.76) $E = C/\Lambda$, $L_0 = TC/\Lambda$, $L_1 = T\Lambda$.

(In fact, we could look at a Legendrian immersion $\iota : \Lambda \to S^*M$, and write the bundles similar to (4.5.1)).

Definition 4.5.15. *The cohomology classes*

(4.5.77) $\mu_h(\Lambda) = \mu_h(E, L_0, L_1) \in H^{4h-3}(\Lambda, \mathbb{R})$

of the bundles (4.5.76) are called the ***Maslov classes of*** Λ ***in*** S^*M.

Here, a remark similar to Remark 5.4.2 can also be added, and we already have an example (Example 4.3.3b)) of such classes.

We want to give a computational machinery for the Maslov classes (4.5.77). We shall see that the geometric structures needed for this computation are induced from T^*M.

Proposition 4.5.16. *Let* J *be the almost complex structure, defined on* T^*M *by Proposition 4.5.3. Then* $-J$ *defines a positive compatible com-*

plex structure of C. *The corresponding hermitian metric is the metric induced by* g *of the same Proposition 4.5.3.*

Proof. Let us begin by noticing that $H/S^*M \subset TS^*M$, where H is the horizontal distribution of T^*M. Indeed, H is defined by (4.5.5), and, on the other hand, it follows easily from (4.5.73) that TS^*M is defined in TT^*M by

(4.5.78) $$\gamma^{\alpha\beta}\xi_\alpha\theta_\beta = 0 .$$

(We use Greek indices as running from 1 to n+1, while the Latin indices will continue to run from 1 to n. So we look at (4.5.5), and other similar formulas, as having Greek indices now.) Furthermore, a comparison of (3.2.25), (4.5.9) and (4.5.13) shows that

(4.5.79) $$A = \gamma^{\alpha\beta}\xi_\alpha X_\beta = \xi_\alpha Y^\alpha$$

and, therefore $A \in H$. But $A \notin C$, hence H and C are transversal distributions on S^*M, and it follows that $K = C \cap H$ is a Lagrangian subbundle of C. Obviously, K and C are transversal to each other in C.

Moreover, it follows from the definition of the metric g of T^*M (Proposition 4.5.3) that K and C are orthogonal distributions. It follows also from 4.5.79 that $A \perp C$, and we shall prove now that $A \perp K$ as well. Indeed, if $Z = \zeta_\alpha Y^\alpha \in K$ we have

$$g(A,Z) = \gamma^{\alpha\beta}\xi_\alpha\zeta_\beta = \eta(Z) = 0 .$$

Hence we have the orthogonal decomposition

(4.5.80) $$TS^*M = K \oplus C \oplus \text{span}\{A\}$$

Furthermore, from (4.5.79) and (4.5.15), we get

(4.5.81) $$JA = \xi_\gamma \frac{\partial}{\partial \xi_\gamma} ,$$

and in view of (4.5.78), JA is normal to TS*M. So we have the orthogonal decomposition

(4.5.82) $$TT^*M/S^*M = K \oplus C \oplus \mathrm{span}\{A\} \oplus \mathrm{span}\{JA\} .$$

Since $A \perp K$, $JA \perp JK$, hence $JK \subset TS^*M$, and in view of (4.5.15) we see that

(4.5.83) $$JK = C .$$

From this relation, and Proposition 4.5.3 we conclude easily that -J is a positive compatible complex structure of the symplectic vector bundle $(C,d\eta)$, and the corresponding hermitian metric in $(C,d\eta)$ is the metric induced by g of T*M. Q.e.d.

Now, let us concentrate on the Legendrian submanifold Λ of S*M. Since J and g are induced by T*M, it is natural to look at Λ as a submanifold of (T^*M,J,g), where we also have the complex connection $\hat{\Gamma}$. Let (e_i) $(i = 1,\dots,n)$ be local g-orthonormal frames tangent to Λ. Then, it is clear that (Je_i,A,JA) are local normal frames of Λ, and we have the following *Gauss-Weingarten equations of* Λ *in* T*M

(4.5.84) $$\hat{\nabla} e_i = \mu_i^j e_j + c_i^j (Je_j) + \kappa_i A + \tau_i (JA) ,$$

(4.5.85) $$\hat{\nabla}(Je_i) = -c_i^j e_j + \mu_i^j (Je_j) - \tau_i A + \kappa_i (JA) ,$$

(4.5.86) $$\hat{\nabla} A = - \sum_{j=1}^{n} \kappa_j e_j + \sum_{j=1}^{n} \tau_j (Je_j) + sA$$

(4.5.87) $$\hat{\nabla}(JA) = - \sum_{j=1}^{n} \tau_j e_j - \sum_{j=1}^{n} \kappa_j (Je_j) + s(JA) ,$$

where the coefficients are 1-forms on Λ, and we used the fact that $\hat{\nabla}$ is a complex metric connection that preserves H, hence $g(\hat{\nabla}A,JA) = 0$ and we have no JA-component in (4.5.86). In these equations, (μ_i^j) are the connection forms of the induced connection of Λ, which is a metric connection (hence $\mu_i^j + \mu_j^i = 0$), and, therefore, it may play the role of the connection θ_1 needed in the computation of the Maslov classes (4.5.77). The coefficients $(c_i^j, \kappa_i, \tau_i)$ define the $\hat{\Gamma}$-second fundamental form of Λ in T*M, and (c_i^j, κ_i) define the second fundamental form of Λ in S*M with respect to the connection induced by $\hat{\Gamma}$ in S*M. (The matrix of this induced connection is the right upper (3,3)-corner of coefficients of (4.5.84)-(4.5.87).) The symmetry relation $c_i^j = c_j^i$ holds by the same proof as for b_i^j in (4.5.37).

Furthermore, if we denote by ξ the local matrix of the connection $\hat{\nabla}$ as it appears in (4.5.84)-(4.5.87), we may compute the corresponding matrix Ξ of curvature forms, and establish corresponding Gauss-Codazzi integrability conditions. Let us agree to denote the entries of Ξ by the same letters as the entries of ξ but in capitals. Then the Gauss-Codazzi conditions appear as

(4.5.88)
$$M_i^j = M'^j_i + c_i^h \wedge c_h^j + \kappa_i \wedge \kappa_j + \tau_i \wedge \tau_j \;,$$
$$C_i^j = (\nabla' c)_i^j - \kappa_i \wedge \tau_j + \tau_i \wedge \kappa_j \;,$$

etc. (we do not write down all the Gauss-Codazzi conditions since we don't need them here), where ∇' is the connection induced in Λ with the matrix (μ_i^j), and $M'^j_i = d\mu_i^j - \mu_i^h \wedge \mu_h^j$ is the inner curvature of the submanifold Λ.

Now, we can prove a result which is similar to Theorem 4,5.9 namely

Theorem 4.5.17. *The Maslov classes* $\mu_h(\Lambda)$ *are expressible in terms of the inner curvature of* Λ, *of the C-component of the* $\hat{\Gamma}$*-second fundamental form* (c_i^j) *of* Λ, *and of its inner covariant derivative. If* $c_i^j = 0$, *all the classes* $\mu_h(\Lambda) = 0$.

Proof. We proceed like in Theorem 4.5.9. We saw already what connection θ_1 should be used, and we have

$$\overset{1}{\pi}{}^j_i = \mu^j_i \tag{4.5.89}$$

Furthermore, it is easy to understand that we may use in the role of θ_0 the connection induced by $\hat{\nabla}$ in the contact distribution C, and whose matrix is the right upper (2,2)-corner of formulas (4.5.84)-(4.5.87). Hence, the same computation as for (4.5.45) yields now

$$\overset{0}{\pi}{}^j_i = \mu^j_i - \sqrt{-1}\, c^j_i \quad , \tag{4.5.90}$$

and, therefore, (4.5.46) is replaced now by

$$\alpha^j_i = \sqrt{-1}\, c^j_i \; . \tag{4.5.91}$$

It follows that $\mu_1(\Lambda)$ is represented by

$$\Delta(\theta_0, \theta_1)c_1 = -\frac{1}{2\pi}\, c^i_i \tag{4.5.92}$$

which proves the theorem for $h=1$.

Finally, like in the case of (4.5.48), we have now using the relations (4.5.88)

$$(\theta_t)^j_i = M'^j_i + (1-t)^2 c^h_i \wedge c^j_h - \sqrt{-1}(1-t)(\nabla' c)^j_i \; , \tag{4.5.93}$$

and Theorem 4.5.17 is proven from this expression of $(\theta_t)^j_i$, and from (4.4.43). Q.e.d.

Corollary 4.5.18. *If* Λ *is a totally geodesic submanifold of* S^*M *with respect to the connection induced by* $\hat{\Gamma}$ *then all* $\mu_h(\Lambda) = 0$.

Corollary 4.5.19. *If* $\hat{\Gamma}$ *induces in* Λ *a flat connection, and if the* C*-component of the* $\hat{\Gamma}$*-second fundamental form of* Λ *is parallel along* Λ, *then* $\mu_h(M)$ *are represented by the forms* m_h *given by formula (4.5.50 i)) where* b^j_i *are replaced by* c^j_i.

Unfortunately, we cannot find a result analogous to Theorem 4.5.11 ii) since the present Gauss-Codazzi conditions (4.5.88) are more complicated. However, we may obtain a result similar to the one given in Proposition 4.5.12. Namely, we have

Proposition 4.5.20. *Let* M *be a flat Riemannian manifold. Let* Λ *be a Legendrian submanifold of* S^*M, *and* H *its mean curvature vector with respect to the riemannian structure induced in* S^*M *by* T^*M. *Then we have*

$$(4.5.94)\qquad \mu_1(\Lambda) = \left[\frac{\kappa}{2\pi}\, i(H')d\eta\right] \quad ,$$

where η *is the contact form of* S^*M, *and* H' *is the orthogonal projection of* H *onto the contact distribution* $\eta = 0$.

Proof. From (4.5.84), and in analogy to (4.5.52), we get

$$(4.5.95)\qquad H = \frac{1}{n}\sum_{i=1}^{n}\left[c_i^j(e_i)(Je_j) + \kappa_i(e_i)A\right] \quad ,$$

and therefore

$$(4.5.96)\qquad H' = \frac{1}{n}\sum_{i=1}^{n} c_i^j(e_i)(Je_j) \ .$$

Now, the same computations done for H in the proof of Proposition 4.5.12 hold for c_i^j instead of b_i^j, which yields

$$(4.5.97)\qquad i(H')d\eta = -\frac{1}{n}\sum_{i=1}^{n} c_i^i \ ,$$

and the requested results follow from (4.5.92). Q.e.d.

Corollary 4.5.21. *If* $H' = 0$, *and particularly if* Λ *is a minimal submanifold of* S^*M *(i.e.,* $H=0$*),* $\mu_1(\Lambda) = 0$.

Remark 4.5.22. *If we have a manifold* M, *and a* ***Lagrangian function*** $L : TM \to \mathbb{R}$, *we get a well known sumplectic structure on* TM [1]. *This* *structure is related to the canonical symplectic structure of* T^*M *by means of the so-called* ***Liouville transformation*** *(which is well defined in both directions if* L *is* ***regular*** [1]. *The computations of the present section can then be transposed to* TM *via the same Liouville transformation.*

Let us add to the computations of the present section the following simple rigidity result

Proposition 4.5.23. *Let* $f : N \times I \to T^*M$ *(or* $f : N \times I \to S^*M$*) be a differentiable mapping such that* $f/N \times \{s\}$ *is a Lagrangian (Legendrian) immersion for every* $s \in I = [0,1]$. *Then* $\mu_h(N \times \{s\}) = \text{const.}$ *in* $H^*(N,\mathbb{R})$.

Proof. By definition, we have $\mu_h(N \times \{s\}) = \mu_h(E_s, L_s, TN)$, where $E_s = f_s^{-1}(TT^*N)$, $L_s = f_s^{-1}(V)$, in the Lagrangian case, and something similar in the Legendrian case. Since the mappings f_s are homotopic, all the pairs $(E_s, L_s)_{s\in I}$ are isomorphic to a fixed pair, say (E_0, L_0), by a differentiable family of isomorphisms $\alpha_s : (E_s, L_s) \to (E_0, L_0)$, and it follows that

$$\mu_h(N \times \{s\}) = \mu_h\big(E_0, L_0, \alpha_s(TN)\big) . \tag{4.5.98}$$

But the classes in the righthandside of (4.5.98) are constant in view of Theorem 4.4.25. Q.e.d.

We shall finish by indicating a topological application of the Maslov classes studied in this section. We consider only the case of Lagrangian submanifolds of T^*M^n but the case of Legendrian submanifolds of S^*M^n can be discussed analogously. Moreover, *all the Lagrangian submanifolds encountered hereafter are assumed to be compact and oriented.*

Definition 4.5.24. *Let* N^n *be an immersed Lagrangian submanifold of* T^*M, *and let* α *be an element of grade* n *in the algebra* $\widehat{WL}_n$ *defined by (4.4.73). Let* $\alpha(N) \in H^n(N,\mathbb{R})$ *be the* n*-dimensional general secondary characteristic class of the bundles (4.5.1), defined by* α. *Then*

$$a(N,\alpha) = \int_N \alpha(N) \tag{4.5.99}$$

is called the ***Maslov number*** *associated to* α.

We remember that by Theorem 4.4.37, α is a product of Pontrjagin and Maslov classes of the corresponding bundles (4.5.1). The application which we have in mind is to the relation of *cylindrical Lagrangian cobordism* introduced by Arnold [4].

Let us consider the cylinder $X = M \times I$, where $I = [0,1]$ is described by a parameter t. Then we have

$$T^*X = T^*M \times I \times \mathbb{R} , \tag{4.5.100}$$

$$\partial T^*X = (T^*M \times \{1\}) \cup (T^*M \times \{0\}) \times \mathbb{R} = (T^*\partial X) \times \mathbb{R} , \tag{4.5.101}$$

where $\mathbb{R}$ is described by another parameter τ.

Let L be a Lagrangian submanifold of the cotangent bundle T^*X which is transversal to ∂T^*X, and such that $\partial L = L \cap \partial T^*X$. For L, we may use local parametric equations like in Proposition 3.3.1. Because of the transversality condition imposed, and using the notation of (3.3.1), we see that these parametric equations are of the form

$$\begin{aligned} x^{\bar{I}} &= x^{\bar{I}}(x^I, \xi_{\bar{I}}, t) , \\ \xi_I &= \xi_I(x^I, \xi_{\bar{I}}, t) , \\ \tau &= \tau(x^I, \xi_{\bar{I}}, t) , \end{aligned} \tag{4.5.102}$$

on neighbourhoods of boundary points of L. It follows that ∂L is an n-dimensional submanifold which projects by (4.5.101) onto an immersed Lagrangian submanifold $\hat{L}$ of $T^*\partial X$.

Definition 4.5.25. [4]. *The manifold* $\hat{L}$ *constructed above is called the* ***Lagrangian boundary*** *of* L. *If* N_a $(a = 0,1)$ *are Lagrangian submanifolds of* T^*M, *and if there exist a Lagrangian submanifold* L *of* T^*X *as described above such that*

$$\hat{L} = N_1 - N_0 \tag{4.5.103}$$

(where, of course, minus stands for reversed orientation) then N_0 *and* N_1 *are said to be* ***cylindrically Lagrangian cobordant manifolds.***

Now, we can prove

Proposition 4.5.26. *The Maslov numbers of a compact oriented manifold of a cotangent bundle are invariant by cylindrical Lagrangian cobordism.*

Proof. If we use the same notation as above, we must prove that (4.5.103) implies

$$a(N_0,\alpha) = a(N_1,\alpha) \tag{4.5.104}$$

for every $\alpha \in \widehat{WL}_n$ of degree n.

Because of (4.5.100), it makes sense to talk of the following vector bundles

$$E = TT^*M/L,\ L_0 = VT^*M/L\ ,\ L_1 = \text{projection of } TL \text{ onto } TT^*M, \tag{4.5.105}$$

where E is a symplectic vector bundle, and L_0, L_1 are Lagrangian subbundles. (As usual, V denotes the tangent bundle of the fibers of T^*M.) Accordingly, we have a well defined closed n-form $\rho(\theta_0,\theta_1)\alpha$ on L (see the definition of $\rho(\theta_0,\theta_1)$ in (4.4.72)) which defines the characteristic class of (E,L_0,L_1) associated to $\alpha \in \widehat{WL}_n$, and

(4.5.106) $$\int_{\partial L} \rho(\theta_0, \theta_1)\alpha = \int_L d\rho(\theta_0, \theta_1)\alpha = 0 .$$

Now, the local equations (4.5.102) of L show that ∂L has a neighbourhood U in L which is diffeomorphic to its projection $\hat{U}$ by (4.5.100) into $T^*M \times T$. This projection sends ∂L onto $\hat{L}$, and it also sends the vector bundles (4.5.105), and their connections, onto similarly defined bundles and connections on $\hat{U}$, such that (4.5.106) is, in fact, a relation which holds on $\hat{L}$. In view of (4.5.103), the relation obtained on $\hat{L}$ actually means

$$0 = \int_{\hat{L}} \rho(\theta_0, \theta_1)\alpha = \int_{N_1} \alpha(N_1) - \int_{N_0} \alpha(N_0) .$$

Q.e.d.

For the study of Lagrangian and Legendrian cobordism we send to [4], [28], [8].

REFERENCES

[1] R. Abraham and J.E. Marsden, Foundations of Mechanics, 2nd. ed. Benjamin/Cummings Co., Reading, Mass. 1978.

[2] V.I. Arnold, Characteristic class entering in quantization conditions. Funkt. Anal. i. Priloz. 1(1) (1967), 1-14 (in Russian). Funct. Anal. Appl. 1 (1967), 1-13 (English transl.).

[3] V.I. Arnold, Mathematical methods of classical mechanics. Nauka, Moskva 1974 (in Russian). Springer, Berlin-New York, 1978 (English transl.).

[4] V.I. Arnold, Lagrange and Legendre cobordism I, II. Funkt. Anal. i. Prilož. 14(3) (1980), 1-13; 14(4) (1980), 8-17 (in Russian). Funct. Anal. Appl. 14 (1980), 167-177; 252-260 (English transl.).

[5] V.I. Arnold, Singularities of systems of rays. Uspekhi Mat. Nauk 38(2) (1983), 77-147 (in Russian). Russian Math. Surveys 38(2) (1983), 87-176 (English transl.).

[6] M. Audin, Classes caractéristiques d'immersions lagrangiennes définies par des variétés de caustiques (d'après V.A. Vassiliev).In: Séminaire Sud Rhodanien de Géométrie I, Géométrie symplectique et de contact (P. Dazord, N. Desolneux Moulis, eds.), Travaux en cours, Hermann, Paris, 1984, 1-16.

[7] M. Audin, Classes caractéristiques Lagrangiennes, Preprint 1986.

[8] M. Audin, Cobordismes d'immersions Lagrangiennes et Legendriennes. Thèse, Univ. Paris-Sud, Orsay, 1986.

[9] D. Bennequin, Caustiques mistiques. In: Séminaire Bourbaki 1984, nr. 634, Astérisque 133-134 (1986), 19-56.

[10] M. Berger, Géométrie 1/actions de groupes, espaces affines et projectifs. Cedic/Nathan, Paris 1977.

[11] I.N. Bernshtein and B.I. Rozenfeld, On characteristic classes of foliations. Funkt. Anal. i Prilož. 6(1) (1972), 68-69 (in Russian). Funct. Anal. Appl. 6 (1972), 60-61 (English transl.).

[12] R.L. Bishop and R.J. Crittenden, Geometry of manifolds. Academic Press, New York 1964.

[13] D. Blair, Contact manifolds in riemannian geometry. Lect. Notes in Math., 509, Springer, Berlin-New York, 1976.

[14] A. Borel, Sur la cohomologie des espaces fibrés principaux et des espaces homogènes de groupe de Lie compacts. Ann. Math. 57 (1953), 115-207.

[15] R. Bott, Lectures on characteristic classes and foliations. In: Lectures on Algebraic and Diff. Top., Lect. Notes in Math. 279, Springer, Berlin-New York, 1972, 1-94.

[16] R. Bott and L.W. Tu, Differential forms in algebraic topology. Springer, Berlin-New York, 1982.

[17] J. Cheeger and J. Simons, Differential characters and geometric invariants. Preprint.

[18] S.S. Chern and J. Simons, Characteristic forms and geometric invariants. Ann. Math. 99 (1974), 48-69.

[19] P. Chernoff and J. Marsden, Properties of infinite dimensional Hamiltonian systems. Lect. Notes in Math. 425, Springer, Berlin-New York 1974.

[20] L.A. Cordero, M. Fernández and A. Gray, Symplectic manifolds with no Kähler structure. Topology 25 (1986), 375-380.

[21] A. Crumeyrolle, Indice d'inertie trilatère et classes de Chern. Studia Scient. Math. Hungarica 13 (1978), 259-271.

[22] P. Dazord, Invariants homotopiques attachés aux fibrés symplectiques. Ann. Inst. Fourier Grenoble, 29(2) (1979), 25-78.

[23] P. Dazord, Sur la géométrie des sous-fibrés et des feuilletages lagrangiens. Ann. Scient. Éc. Norm. Sup. 13 (1981), 465-480.

[24] P. Dazord, Feuilletages et mécanique hamiltonienne. Publ. Dept. Math. Lyon, Exp. 4, 1983 (with an Erratum in same Publ. 1984).

[25] R. Deheuvels, Formes quadratiques et groupes classiques. Presse Univ. France, Paris 1981.

[26] Th. Delzant, Feuilletages de variétés symplectiques. C.R. Acad. Sci. Paris 300 I (1985), 201-204.

[27] D. Mc. Duff, Examples of simply-connected symplectic manifolds. J. Diff. Geometry 20 (1984), 267-278.

[28] J. Eliashberg, Cobordisme des solutions de relations différentielles. In: Séminaire Sud-Rhodanien de Géométrie I. Géométrie symplectique et de contact (P. Dazord, N. Desolneux Moulis, eds.), Travaux en cours, Hermann, Paris, 1984, 17-32.

[29] D.B. Fuks, Maslov-Arnold characteristic classes. Dokl. Akad. Nauk. SSSR 178(2) (1968), 303-306. (in Russian). Soviet Math. Dokl. 9 (1968), 96-99. (English transl.)

[30] C. Godbillon and J. Vey, Un invariant des feuilletages de codimension un. C.R. Acad. Sci. Paris 273 A (1971), 92-95.

[31] S.I. Goldberg, Curvature and homology. Academic Press, New York 1962.

[32] M.J. Gotay, On coisotropic imbeddings of presymplectic manifolds. Proceedings American Math. Soc. 84 (1982), 111-114.

[33] A. Gray and L.M. Hervella, The sixteen classes of almost hermitian manifolds. Annali Mat. Pura Appl. 123 (1980), 35-58.

[34] W. Greub, S. Halperin, R. Vanstone, Connections, Curvature and Cohomology I, II, III. Academic Press, New York, 1972, 1973, 1976.

[35] M.L. Gromov, A topological technique for the construction of solutions of differential equations and inequalities. Internat. Congress of Math. Nice 1970, Vol. 2, 221-225, Gauthier-Villars, Paris 1971.

[36] M.L. Gromov, Pseudo holomorphic curves in symplectic manifolds. Inventiones Math. 82 (1985), 307-347.

[37] V. Guillemin and Sh. Sternberg, Geometric asymptotics. Math. Surveys 14, American Math. Soc., Providence, R.I. 1977.

[38] A. Haefliger, Sur les classes caractéristiques des feuilletages. In: Sém. Bourbaki 1971/72, exp. 412, Lect. Notes in Math. 317, Springer-Verlag, Berlin-New York 1973, 239-260.

[39] J.L. Heitsch, Deformations of secondary characteristic classes. Topology 12 (1973), 381-388.

[40] M.W. Hirsch, Differential topology. Springer, Berlin-New York 1976.

[41] L. Hörmander, Fourier integral operators I. Acta Math. 127 (1971), 79-183.

[42] D. Husemoller, Fibre bundles, 2nd ed. Springer, Berlin-New York 1975.

[43] F.W. Kamber and Ph. Tondeur, Foliated bundles and characteristic classes. Lect. Notes in Math. 493, Springer, Berlin-New York, 1975.

[44] J.B. Keller, Corrected Bohr-Sommerfeld quantum conditions for nonseparable systems. Ann. Physics, 4 (1958), 180-188.

[45] S. Kobayashi and K. Nomizu, Foundations of differential geometry I, II. Intersc. Publ., New York 1963, 1969.

[46] S. Lang, Algebra. Addison-Wesley Co., Reading, Mass. 1965.

[47] S. Lang, Differentiable manifolds, 2nd ed. Addison-Wesley Co. Reading, Mass. 1972.

[48] J. A. Lees, On the classification of Lagrange immersions. Duke Math. J. 43 (1976), 217-224.

[49] D. Lehmann, Classes caractéristiques et J-connexité des espaces de connexions. Ann. Inst. Fourier Grenoble, 24 (3) (1974), 267-306.

[50] D. Lehmann, Résidus des connexions à singularités et classes caractéristiques. Ann. Inst. Fourier Grenoble, 31(1) (1981), 83-98.

[51] J. Leray, Analyse lagrangienne et mécanique quantique. IRMA Univ. Strasbourg, Vol. 25, Strasbourg 1978.

[52] P. Libermann, Sur le problème d'équivalence de certaines structures infinitésimales. Annali Mat. Pura Appl. 36 (1954), 27-120.

[53] P. Libermann, Sous-variétés et feuilletages symplectiquement réguliers. In: Symplectic geometry (Crumeyrolle and Grifone eds.), Research Notes in Math. 80, Pitman, London 1983, 81-106.

[54] P. Libermann, Problèmes d'équivalence et geométrie symplectique, Astérisque 107-108 (1983), 43-68.

[55] A. Lichnerowicz, Generalized foliations and local Lie algebras of Kirillov. In: Differential Geometry (L.A. Cordero, ed.) Research Notes in Math. 131, Pitman, London 1985, 198-210.

[56] Ch. M. Marle, Sous-variétés de rang constant d'une variété symplectique, Astérisque 107-108 (1983), 69-86.

[57] V.P. Maslov, Theory of perturbations and asymptotic methods. Moskow Univ. Publ. House, Moskwa 1965 (in Russian). Dunod Paris 1972 (French transl.)

[58] V.P. Maslov and M.V. Fedoriuk, Semi-classical approximation in quantum mechanics. Nauka, Moskwa 1976 (in Russian). D. Reidel Co., Dordrecht 1981 (English transl.).

[59] Y. Matsushima, Differentiable manifolds. M. Dekker, Inc. New York 1972.

[60] J. Milnor and J.D. Stasheff, Characteristic classes. Princeton Univ. Press, Princeton N.J. 1974.

[61] A.S. Misčenko, B.Iu. Sternin and V.E. Shatalov, Lagrangian submanifolds and the method of the canonical operator. Nauka, Moskwa 1978 (in Russian).

[62] J.M. Morvan, Quelques invariants topologiques en géométrie symplectique. Ann. Inst. H. Poincaré, A, 38 (1983), 349-370.

[63] J.M. Morvan, Une nouvelle obstruction topologique à la transversalité de deux fibrés lagrangiens. Sém. Géom. Diff. 1982-1983, Exp. 9. Univ. du Languedoc, Montpellier, France.

[64] J.M. Morvan, Invariants locaux et globaux en géométrie symplectique. Sém. Géom. Diff. 1982-1983, Exp. 10, Univ. du Languedoc, Montpellier, France.

[65] J.M. Morvan, Connexions et feuilletages en géométrie symplectique. C.R. Acad. Sci. Paris 296 I (1983), 765-768.

[66] J.M. Morvan and L. Niglio, Classes caractéristiques des couples de sous-fibrés lagrangiens. Ann. Inst. Fourier Grenoble 36 (1986), 193-209.

[67] J. Moser, On the volume elements on a manifold. Trans. American Math. Soc. 120 (1965), 286-294.

[68] M.S. Narasimhan and S. Ramanan, Existence of universal connections I, II. American J. Math. 83 (1961), 563-752; 85 (1963), 223-231.

[69] R.S. Palais, A global formulation of the Lie theory of transformation groups. Memoirs American Math. Soc. 22, Providence, R.I. 1957.

[70] B.L. Reinhart, Differential geometry of foliations, Springer, Berlin-New York 1983.

[71] V.A. Rochlin and D.B. Fuks, An introductory course in topology. Geometric chapters. Nauka, Moskwa 1977 (in Russian).

[72] C.L. Siegel, Symplectic geometry. Academic Press, New York 1964.

[73] J. Simons, Characteristic forms and transgression II. Characters associated to a connection. Preprint.

[74] J.M. Souriau, Construction explicite de l'indice de Maslov. Applications. Group theoretical methods in physics (Fourth Internat. Colloq., Nijmegen 1975). Lect. Notes in Physics 50, Springer, Berlin-New York, 1976, 117-148.

[75] E.H. Spanier, Algebraic topology. Mc Graw-Hill, New York 1966.

[76] N. Steenrod, The topology of fibre bundles. Princeton Univ. Press, Princeton N.J. 1951.

[77] Sh. Sternberg, On the role of field theories in our physical conception of geometry. In: Proc. Bonn Conf. on Diff. Geom. Methods in Physics 1977. Lect. Notes in Math. 676, Springer, Berlin-New York 1978, 1-80.

[78] W.P. Thurston, Some simple examples of symplectic manifolds. Proc. American Math. Soc. 55 (1976), 467-468.

[79] Ph. Tondeur, Structure presque Kählérienne naturelle sur le fibré des vecteurs covariants d'une variété riemanienne. C.R. Acad. Sci. Paris 254 (1962), 407-408.

[80] I. Vaisman, Cohomology and differential forms. M. Dekker, Inc., New York 1973.

[81] I. Vaisman, Geometric quantization of presymplectic manifolds, Monatshefte für Math. 96 (1983), 293-310.

[82] I. Vaisman, Locally conformal symplectic manifolds. Internat. J. of Math. and Math. Sci. 8 (1985), 521-536.

[83] I. Vaisman, Lagrangian foliations and characteristic classes. In: Differential Geometry (L.A. Cordero, ed.) Research Notes in Math. 131, Pitman, Inc., London 1985, 245-256.

[84] I. Vaisman, The Maslov class of some Legendre submanifolds. J. of Geometry and Physics 3 (1986).

[85] V.A. Vassiliev, Characteristic classes of Lagrangian and Legendrian immersions dual to singularities of caustics and wave fronts. Funkt. Anal. i Prilož. 15(3) (1981), 10-22 (in Russian). Funct. Anal. Appl. 15 (1981), 164-172.(English transl.)

[86] A. Weinstein, Symplectic manifolds and their Lagrangian submanifolds. Advances in Math. 6 (1971), 329-346.

[87] A. Weinstein, Lectures on symplectic manifolds. CBMS Conf. Series 29, American Math. Soc. Providence, R.I., 1977.

[88] A. Weinstein, The local structure of Poisson manifolds, J. Diff. Geometry 18 (1983), 523-557.

[89] K. Yano, On a structure defined by a field of type (1,1) satisfying $f^3+f = 0$. Tensor 14 (1963), 99-109.

[90] K. Yano and S. Ishihara, Tangent and cotangent bundles. M. Dekker, Inc., New York 1973.

INDEX

Progress in Mathematics

Edited by:

J. Oesterlé
Departement des Mathematiques
Université de Paris VI
4, Place Jussieu
75230 Paris Cedex 05
France

A. Weinstein
Department of Mathematics
University of California
Berkeley, CA 94720
U.S.A.

Progress in Mathematics is a series of books intended for professional mathematicians and scientists, encompassing all areas of pure mathematics. This distinguished series, which began in 1979, includes authored monographs and edited collections of papers on important research developments as well as expositions of particular subject areas.

All books in the series are "camera-ready", that is they are photographically reproduced and printed directly from a final-edited manuscript that has been prepared by the author. Manuscripts should be no less than 100 and preferably no more than 500 pages.

Proposals should be sent directly to the editors or to: Birkhauser Boston, 675 Massachusetts Avenue, Suite 601, Cambridge, MA 02139, U.S.A.

1 GROSS. Quadratic Forms in Infinite-Dimensional Vector Spaces
2 PHAM. Singularités des Systèmes Différentiels de Gauss-Manin
3 OKONEK/SCHNEIDER/SPINDLER. Vector Bundles on Complex Projective Spaces
4 AUPETIT. Complex Approximation, Proceedings, Quebec, Canada, July 3–8, 1978
5 HELGASON. The Radon Transform
6 LION/VERGNE. The Weil Representation, Maslov Index and Theta Series
7 HIRSCHOWITZ. Vector Bundles and Differential Equations Proceedings, Nice, France, June 12–17, 1979
8 GUCKENHEIMER/MOSER/NEWHOUSE. Dynamical Systems, C.I.M.E. Lectures, Bressanone, Italy, June 1978
9 SPRINGER. Linear Algebraic Groups
10 KATOK. Ergodic Theory and Dynamical Systems I
11 BALSLEV. 18th Scandinavian Congress of Mathematicians, Aarhus, Denmark, 1980
12 BERTIN. Séminaire de Théorie des Nombres, Paris 1979–80
13 HELGASON. Topics in Harmonic Analysis on Homogeneous Spaces
14 HANO/MARIMOTO/MURAKAMI/OKAMOTO/OZEKI. Manifolds and Lie Groups: Papers in Honor of Yozo Matsushima
15 VOGAN. Representations of Real Reductive Lie Groups
16 GRIFFITHS/MORGAN. Rational Homotopy Theory and Differential Forms
17 VOVSI. Triangular Products of Group Representations and Their Applications
18 FRESNEL/VAN DER PUT. Géometrie Analytique Rigide et Applications
19 ODA. Periods of Hilbert Modular Surfaces
20 STEVENS. Arithmetic on Modular Curves
21 KATOK. Ergodic Theory and Dynamical Systems II
22 BERTIN. Séminaire de Théorie des Nombres, Paris 1980–81
23 WEIL. Adeles and Algebraic Groups

24 LE BARZ/HERVIER. Enumerative Geometry and Classical Algebraic Geometry
25 GRIFFITHS. Exterior Differential Systems and the Calculus of Variations
26 KOBLITZ. Number Theory Related to Fermat's Last Theorem
27 BROCKETT/MILLMAN/SUSSMAN. Differential Geometric Control Theory
28 MUMFORD. Tata Lectures on Theta I
29 FRIEDMAN/MORRISON. Birational Geometry of Degenerations
30 YANO/KON. CR Submanifolds of Kaehlerian and Sasakian Manifolds
31 BERTRAND/WALDSCHMIDT. Approximations Diophantiennes et Nombres Transcendants
32 BOOKS/GRAY/REINHART. Differential Geometry
33 ZUILY. Uniqueness and Non-Uniqueness in the Cauchy Problem
34 KASHIWARA. Systems of Microdifferential Equations
35 ARTIN/TATE. Arithmetic and Geometry: Papers Dedicated to I.R. Shafarevich on the Occasion of His Sixtieth Birthday, Vol. I
36 ARTIN/TATE. Arithmetic and Geometry: Papers Dedicated to I.R. Shafarevich on the Occasion of His Sixtieth Birthday, Vol. II
37 DE MONVEL. Mathématique et Physique
38 BERTIN. Séminaire de Théorie des Nombres, Paris 1981–82
39 UENO. Classification of Algebraic and Analytic Manifolds
40 TROMBI. Representation Theory of Reductive Groups
41 STANELY. Combinatories and Commutative Algebra
42 JOUANOLOU. Théorèmes de Bertini et Applications
43 MUMFORD. Tata Lectures on Theta II
44 KAC. Infinite Dimensional Lie Algebras
45 BISMUT. Large Deviations and the Malliavin Calculus
46 SATAKE/MORITA. Automorphic Forms of Several Variables Taniguchi Symposium, Katata, 1983
47 TATE. Les Conjectures de Stark sur les Fonctions L d'Artin en s = 0
48 FRÖHLICH. Classgroups and Hermitian Modules
49 SCHLICHTKRULL. Hyperfunctions and Harmonic Analysis on Symmetric Spaces
50 BOREL, ET AL. Intersection Cohomology
51 BERTIN/GOLDSTEIN. Séminaire de Théoire des Nombres. Paris 1982–83
52 GASQUI/GOLDSCHMIDT. Déformations Infinitesimales des Structures Conformes Plates
53 LAURENT. Théorie de la Deuxième Microlocalisation dans le Domaine Complèxe
54 VERDIER/LE POTIER. Module des Fibres Stables sur les Courbes Algébriques Notes de l'Ecole Normale Supérieure, Printemps, 1983
55 EICHLER/ZAGIER. The Theory of Jacobi Forms
56 SHIFFMAN/SOMMESE. Vanishing Theorems on Complex Manifolds
57 RIESEL. Prime Numbers and Computer Methods for Factorization
58 HELFFER/NOURRIGAT. Hypoellipticité Maximale pour des Opérateurs Polynomes de Champs de Vecteurs
59 GOLDSTEIN. Séminaric de Théorie des Nombres, Paris 1983–84
60 PROCESI. Geometry Today: Giornate Di Geometria, Roma. 1984
61 BALLMANN/GROMOV/SCHROEDER. Manifolds of Nonpositive Curvature
62 GUILLOU/MARIN. A la Recherche de la Topologie Perdue
63 GOLDSTEIN. Séminaire de Théorie des Nombres, Paris 1984–85
64 MYUNG. Malcev-Admissible Algebras
65 GRUBB. Functional Calculus of Pseudo-Differential Boundary Problems

66 Cassou-Noguès/Taylor. Elliptic Functions and Rings and Integers

67 Howe. Discrete Groups in Geometry and Analysis: Papers in Honor of G.D., Mostow on His Sixtieth Birthday

68 Robert. Antour de L'Approximation Semi-Classique

69 Faraut/Harzallah. Deux Cours d'Analyse

70 Adolphson/Conrey/Gosh/Yager. Number Theory and Diophantine Problems: Proceedings of a Conference at Oklahoma State University

GPSR Compliance

The European Union's (EU) General Product Safety Regulation (GPSR) is a set of rules that requires consumer products to be safe and our obligations to ensure this.

If you have any concerns about our products, you can contact us on ProductSafety@springernature.com

In case Publisher is established outside the EU, the EU authorized representative is:

Springer Nature Customer Service Center GmbH
Europaplatz 3
69115 Heidelberg, Germany

Batch number: 10577876

Printed by Printforce, the Netherlands